ECONOMICS OF TRANSFRONTIER POLLUTION

ORGANISATION FOR ECONOMIC CO-OPERATION AND DEVELOPMENT

The Organisation for Economic Co-operation and Development (OECD) was set up under a Convention signed in Paris on 14th December, 1960, which provides that the OECD shall promote policies designed:

- to achieve the highest sustainable economic growth and employment and a rising standard of living in Member countries, while maintaining financial stability, and thus to contribute to the development of the world economy;
- to contribute to sound economic expansion in Member as well as non-member countries in the process of economic development;
- to contribute to the expansion of world trade on a multilateral, non-discriminatory basis in accordance with international obligations.

The Members of OECD are Australia, Austria, Belgium, Canada, Denmark, Finland, France, the Federal Republic of Germany, Greece, Iceland, Ireland, Italy, Japan, Luxembourg, the Netherlands, New Zealand, Norway, Portugal, Spain, Sweden, Switzerland, Turkey, the United Kingdom and the United States.

* *
*

TABLE OF CONTENTS

FOREWORD

The OECD recently published a book on Problems in Transfrontier Pollution. Although it was interdisciplinary in character, about 80 per cent of the contributions were written by economists. The present publication is wholly of an economic character and it includes papers discussed in the OECD Environment Committee's Sub-Committee of Economic Experts; these papers were prepared by delegates, by members of the OECD Secretariat and by consultants. A critical reader might therefore ask why there is such a continuous effort directed towards transfrontier pollution, and why so much emphasis is placed on the economic aspects ?

The first question is easily answered. Environmental problems present two aspects which are truly international in character - their impact on international trade and investment, and the exchange of pollution between countries (transfrontier pollution). It is therefore clear that the OECD should also contribute actively to the solution of problems of the latter type.

The second question is concerned with the economic nature of the papers presented here. From a general stand point, let us first quote a passage by Herfindahl and Kneese who, introducing a book on environmental quality wrote: "We feel, however - and the reader should be warned that we are economists - that the discipline of economists is central to progress on these problems, for it is economics alone that can formulate these problems in the terms to which they must finally be reduced, namely the balancing of our varied desires in these matters against the costs of satisfying them in various degrees" (1).

More specificially, however, the sizeable effort of the Sub-Committee of Economic Experts devoted to transfrontier pollution can be judged by reference to the fact that if a map of economic knowledge had been drawn three years ago, before the OECD's contribution to this subject, the region representing transfrontier pollution would have been described by the words often found on mediaeval maps : "hic sunt leones" (2).

This publication will therefore serve as a useful "box of tools" for understanding and solving transfrontier pollution problems. The main questions dealt with are: General principles (e.g. the polluter-pays principle (PPP) and the non-discrimination principle), instruments (including an original application of reciprocal taxation to transfrontier pollution), and institutions.

Given the state of the art, contributions are mainly of a theoretical character but, contrary to what first impressions might suggest, this certainly does not at all mean that they are irrelevant for decision-making. It is quite evident that sound action must be based on an appropriate conceptual background. This will be clear, for example, from the paper by L. Ruff which explains the role of the polluter-pays-principle in transfrontier pollution problems and the application of the non-discrimination principle to such problems.

---

1) See Herfindahl O.C., Kneese A.V., Quality of the Environment: An Economic Approach to Some Problems in Using Land, Water, and Air. Baltimore, The John Hopkins Press for the Future, Second Edition, 1967, p. vi.

2) For further details on this point see page 178 of Professor Scott's paper in this volume.

Having thus established the economic framework as a first approximation for the understanding of transfrontier pollution, the Sub-Committee of Economic Experts welcomed the creation of an interdisciplinary Transfrontiers Pollution Group which will pursue the difficult task of advising Member countries on the best possible ways of dealing with a somewhat delicate question.

E. Gerelli

Chairman of the
Sub-Committee of Economic Experts

Part One

GENERAL PRINCIPLES

THE ECONOMICS OF TRANSNATIONAL POLLUTION

by

Larry E. Ruff
Environmental Protection Agency (1)
United States

1) The views expressed in this paper are the technical views of the author, as a member of the Sub-Committee of Economic Experts of the Environment Committee, OECD, and do not necessarily reflect the views of the Environmental Protection Agency or of the United States Government.

This paper presents some simple observations on the economics of transfrontier pollution problems. The economic logic of the Polluter Pays Principle (PPP) is discussed and shown not to require or prohibit any payments among governments in transfrontier pollution problems. Then, the transfrontier problem is treated as a simple problem involving externalities among individuals, where the economic principles of solution are well known. The potential role of economists in dealing with such problems is outlined, and it is argued that the Sub-Committee of Economic Experts should seek to play an analogous role in the transfrontier pollution problem.

THE POLLUTER PAYS PRINCIPLE AMONG NATIONS

The Polluter Pays Principle (PPP), as adopted by the Council of the OECD (1), is intended to be purely an efficiency principle. As adopted it is incomplete as an efficiency principle, because it requires only that control costs and not residual damages be paid; but it is clearly not intended to be an equity principle: i.e., it is not suggested in order to "punish" the polluters. Rather, it is intended to incorporate total costs into product prices so that trade flows will be based on true economic scarcity conditions.

As Professor Beckermann has pointed out in his excellent paper on the polluter pays principle (1), the correct price adjustments theoretically can be accomplished with a set of subsidies for pollution control as well as with a set of charges or regulations. The difficulty with such subsidies is that, in practice, they invariably attract entry into the polluting activities, force the prices of the pollution-intensive products below their true social costs, and hence lead to economic distortions. Therefore, the PPP has been suggested as a guide to policy making in the real world, urging that subsidies not be paid to polluters.

In the case of transfrontier pollution, however, the same arguments against payments to the polluting nation do not apply. If country A is polluting country B to an inefficient degree, and country B makes a payment to the government of country A to induce A to restrict its polluters, this need not lead to international inefficiency in production or trade patterns. So long as country A takes action to restrict its own polluters in a way which imposes the external costs on the polluters themselves, any transfer of funds between the two governments has no economic efficiency effects. If B bribes A, then the citizens of country B will be poorer and the citizens of country A will be richer than if A compensated B; but this is purely a wealth effect and is perfectly consistent with economic efficiency in trade. Therefore, the PPP, regarded as an economic efficiency principle, cannot be interpreted to require or prohibit any sort of transfers among governments in transfrontier pollution problems.

TRANSFRONTIER POLLUTION AS A PROBLEM AMONG INDIVIDUALS

If we regard individual nations as sovereign, with power to control their own citizens in any way they choose, then the transfrontier pollution problem is no different from an ordinary problem of externalities among a small group of individuals. The economic principles of a solution to this problem are simple and well known; depending on who has the "rights" to the environmental resource being polluted, either the polluter compensates the pollutee for damages done or the pollutee bribes the polluter to stop his pollution. In either case, the costs of the pollution are internalised and an efficient level of pollution will be obtained. To accomplish this efficient solution it is necessary to (1) establish the property

1) See The Polluting Pays Principle, OECD, 1975.

rights clearly, (2) estimate damages caused to the polluted country in order to determine how much compensation or bribe should be paid, and (3) establish some market-type institution in which the payments can be made and the rights exchanged.

In principle, there is no reason that this approach cannot work for solving transfrontier problems. The determination of the relevant rights is not, strictly speaking, an economic problem, although economists can make suggestions which may make it easier to come to agreement on the matter; for example, we can suggest that strict adherence to the PPP among governments is not required for economic efficiency, thereby possibly reducing the reluctance of some governments to negotiate on the question of rights. As economists we can help governments estimate the costs of pollution damages so that they may more rationally make decisions regarding the transfer of these rights; and we may make suggestions about market institutions in which these rights can be transferred freely and easily in response to changing circumstances.

As a practical matter, it may be that the most important role economists can play in the transfrontier pollution problem is to suggest institutional arrangements and mechanisms in which environmental rights can be exchanged freely and unemotionally, once they have been established by negotiation. One reason governments are reluctant to negotiate about environmental rights is that they have no clear idea how the rights will be used once they have been established. For example, if nation A is polluting nation B unilaterally, nation B typically will be reluctant to agree that A has any rights to such pollution because there exist few institutional arrangements by which A can prevent B from polluting beyond the agreed upon level. Similarly, A may hesitate to agree that its rights to pollute its neighbour are limited for fear that the agreed upon number of rights will become inflexible and non-negotiable limits in the future, even though objective conditions may change. Both governments might be more willing to negotiate concerning the allocation of environmental rights if there existed mechanisms which promised to allow equitable and efficient exchange and protection of these rights in the future. Therefore, if economists can suggest such institutions, they may also help solve the problem of establishing the rights initially.

There are, of course, virtually an infinite number of conceivable market-type institutions in which environmental rights can be exchanged among nations. We have only begun to articulate a few of the possibilities and then in an unsystematic and imprecise manner. Without presuming to have thought these through completely or claiming that they are the best or only possibilities, the following suggestive examples are put forward for discussion:

1. In the case of downstream water pollution where nation A pollutes nation B, A and B could agree upon a "normal" quality of the water entering B and upon a "price" for higher or lower quality. Then, if the quality of the water is worse over some specified time period, A pays B the appropriate price; similarly if the water is cleaner than the normal level, B pays A the appropriate amount. Nation A controls its polluter as it pleases, presumably consistent with the PPP. B could negotiate a similar arrangement with a downstream nation C. At any time, the nations may renegotiate about the "normal" levels, making whatever side payments or political concessions they deem appropriate for obtaining the re-allocation of rights.

2. In the case where many nations pollute a common resource which they all use equally, such as a border lake, the nations could agree on a common effluent charge imposed on all their individual polluters, the revenues to go into a common fund, which is then distributed back to the member governments on the basis of some predetermined formula. So long as the refunds do not go back to the polluters themselves this can be an economically efficient solution, perfectly consistent with the PPP; and the existence of the fund to distribute can be a useful device for buying the agreement of reluctant nations.

Variations on these mechanisms and wholly new ones can be developed by those more imaginative than myself. But, it should be clear that there are many possibilities and we should not foreclose any of them without careful consideration.

CONCLUSIONS

Transfrontier pollution problems are diverse, complex, and contain much more than purely economic considerations. However, there is an important role to be played by economists in suggesting a diversity of possible economic institutions for solution of these problems, by stressing that these problems can be analysed in economic terms, and by separating clearly the political-legal questions from the economic questions.

A NOTE ON SOME PROBLEMS OF TRANSFRONTIER POLLUTION (1)

by

Judith Marquand

Department of the Environment, London
United Kingdom

---

1) This note was prepared as a contribution to the discussion on transfrontier pollution and should not be read as providing any indication of United Kingdom policy.

INTRODUCTION

Discussion of three papers on transfrontier pollution by the Sub-Committee of Economic Experts after the Seminar on transfrontier pollution made it clear that the nature of the problem had not been clearly defined. The purpose of this note is to attempt to clarify some aspects of the problem, and to suggest a possible framework which might lead to procedures for the consideration of particular cases of various types.

The note is in three parts. First, the extent of the guidance provided by the "polluter pays principle" is considered. Secondly, the situation if there is no agreed procedure for examining transfrontier problems is examined. Thirdly, the nature of procedures compatible with international law, the Principles adopted at Stockholm, and the OECD Convention (1960) is examined. A pair of principles is tentatively suggested for discussion. It is possible that these fulfil the theoretical requirements of the problem whilst being in accordance with the body of international law and international agreements to which the OECD countries subscribe.

TRANSFRONTIER POLLUTION AND THE "POLLUTER PAYS PRINCIPLE"

There can be circumstances in which there is no problem of principle in applying the "polluter pays principle" in cases of transfrontier pollution. These arise where the states with jurisdiction over the regions involved agree on common standards to be attained in all of them. If these standards take the form of common emission standards, then all that is required of each state is to enforce compliance with these standards in accordance with the "polluter pays principle". If they take the form of common environmental quality standards, each state must then develop suitable combinations of standards for emissions (some of which may apply to products such as vehicles, while others will apply to emissions from stationary installations) so as to achieve these quality standards, and enforce the resulting emission standards in accordance with the "polluter pays principle". This is the situation envisaged in the Scott/Bramsen paper on guiding principles (1). But such a solution assumes away all the more difficult parts of the whole transfrontier pollution problem. Adoption of the "polluter pays principle" does not imply adoption of common standards; allowance is made explicitly for the different values and different opportunity costs in different countries. In most circumstances it would not be appropriate to expect a poor country to choose to devote as high a proportion of its scarce resources to abating pollution as does its rich neighbour. But the adoption of common standards would often imply a higher proportion of resources in the poor country devoted to pollution abatement than in the rich country. Aside from any considerations of injustice, this is scarcely a situation where agreement between rich and poor could be exptected to be reached.

It may be objected that in an OECD context, at any rate, there are likely to be few transfrontier pollution problems where the countries concerned are at widely differing stages of development or have widely different sets of values. But OECD countries are not all equally rich, and transfrontier pollution problems are not only those common resource problems where countries are contiguous at least to the resource, if not to each other. Transfrontier pollution problems include all cases where one country concerned with a common resources or upstream/downstream transfrontier pollution problem is free to resolve it by deciding to adopt common standards of one sort or another if it so wishes, but to prescribe such a procedure is an inappropriate way of dealing with transfrontier pollution problems in general.

---

1) See papers by Bramsen, Scott and Muraro in Problems of Transfrontier Pollution, OECD, 1974.

Even without agreement on common standards, if countries are prepared to agree on how to determine an optimum point to which pollution should be abated, then the "polluter pays principle" can be used to determine the allocation of control costs between discharging countries in order to attain it. Provided there is international agreement as to the marginal control costs of each country, and to the marginal costs of the damage incurred in each country, then the whole system becomes determinate. Joint pollution between the countries concerned, if they act "rationally" should be abated to the point where their combined marginal damage costs equalled their combined marginal control costs, and the abatement would be allocated between countries according to the marginal control cost curves of each one and the combined marginal damage cost curve. The international optimum would be achieved as a consequence of equating these curves for each country. But the existence of the determinate international optimum point does depend crucially upon agreement between the countries concerned that international resources are in some sense internationally owned, and upon agreement between the countries as to the marginal damage cost curve and marginal control cost curve of each one. Without such agreement, the curves could not be combined, even in principle, and there would be no agreed point to which pollution should be abated.

However, if it is assumed that international damage and cost functions can be drawn up, at least in principle, then the common resource problem would be solved immediately by application of the "polluter pays principle".

Further steps might still be required where the transfrontier pollution problem was an upstream/downstream one. Consider two countries, A and B. B is downstream from A. We have assumed that A accepts B's valuation of the damage which A causes in B, and that both A and A abate pollution to the point where their marginal control costs are equal to each other and to the marginal damage costs in B (and in A). The question which still arises is whether A should ever pay B, or vice versa, for some of the abatement which is undertaken. Strictly, the model implies that no transfrontier payments as such need ever be made. Each country abates according to its own internationally agreed marginal control cost curve, and it is not directly relevant which country receives the benefit of this abatement. It may be, of course, that the most cost-effective way for B to achieve a certain abatement of pollution is for it to be undertaken at a location in A. This might appear as a payment from B to A. Technical considerations alone would not lead to this situation. The circumstances under which it would arise would be where A's marginal control costs were in some sense valued more highly than a similar expenditure by B. But provided all the necessary assumptions hold, we can say that the upstream/downstream problem is solved in principle by international agreement as to each country's marginal control cost functions and marginal damage cost function, hence determining an optimum. The "polluter pays principle" can then be applied to attain this point. Where the downstream country values a given expenditure on control, to be undertaken within the upstream country, as less costly than the value which the upstream-country would attach to the same expenditure if it incurred it itself, attainment of the optimum by applying the "polluter pays principle" may entail a payment by the downstream country to the upstream country.

The crucial assumption required for the optimum point to be determinate is that countries agree each others' marginal control cost curves and damage cost curves. Where each country agrees only to respect the marginal damage cost function and marginal control cost functions of the others, but where the different countries do not agree that the cost functions can be combined, the optimum is indeterminate. An optimal range may be said to be defined by using each country's set of values in turn and hence discovering where each country's view of where the international optimum would lie if other countries shared its set of values. Consider what is involved if a determinate optimum is to be achieved, so that each country has to agree the others' costs and damages. For if each country, using its own set of values, took account of all the opportunity costs implicit in abating pollution - the loss of employment, all the repercussions on growth, etc. - then the marginal control costs curve would vary from country to country and from region to region, even if technical factors and the costs of those inputs directly associated with the control process were identical. For example, if opportunity costs were taken fully into account, the marginal control costs for a given abatement of pollution would often be much higher for a poor country or indeed for any country with an unemployment problem, than for a country where few resources were idle and where public preferences were relatively stronger for a cleaner environment and less strong for more goods. The poor country would tend to want to abate less than the rich country. Similarly, different countries are likely to value differently the damage which pollution causes them. It is not possible to draw general conclusions about the likely variations in valuation of damage. For example, where river pollution reduces the number of fish, this will be differently valued, ceteris paribus, where the fish are a staple item of diet than where all that is at stake is the loss of fishing for recreation. Yet the country where fishing is simply a recreation may be a very rich country which values its leisure highly; in monetary terms it may be prepared to spend more to preserve its recreation than a poor country to preserve its food.

But any solution, either in the common resource case or in the upstream/downstream case, depends on agreement by all the countries concerned on what are the marginal control cost curves and marginal damage cost curves which each country has. If such agreement is lacking, then there can be no agreement that each country is contributing its "fair" share of pollution abatement in any transfrontier pollution case, whether it be the common resources or the upstream/downstream kind. The problems to be overcome in reaching such agreement are formidable. The marginal control cost curves and damage cost curves for each country would have to be estimated explicitly and then the curves for each country would have to be agreed internationally.

If an attempt were made to apply this approach, the first task would be for OECD or whatever other international organisation was relevant to draw up guideline procedures for the estimation of the appropriate marginal control cost and marginal damage cost curves and to persuade the member nations to subscribe to them.

Even if data to estimate the appropriate curves were thought to exist and even though subscription to the "polluter pays principle" might be taken to imply subscription to the principle of such an approach, it cannot be assumed that it would be easy to agree guidelines. For agreeing guidelines would be tantamount to agreeing an international system for valuing goods and services - yet a major advantage of the "polluter pays principle" is that it allows individual countries to make their own valuations (including nil valuations). Hence the "polluter pays principle" can only be invoked internationally in a way which settles transfrontier pollution problems by the abrogation of the right expressly allowed for individual nations to determine their own valuations. If this right is to be upheld, it follows that some additional principle or principles must be invoked if solutions to transfrontier pollution problems compatible with the "polluter pays principle" are to be found.

An analysis in terms of rights does not seem to overcome this difficulty. In the common resources case, even if each country is prepared to pay for the right to a certain amount of "clean" environment, implying an abatement of its own pollution up to the point where its marginal control cost curve cuts its marginal damage cost curve, it is again only if there is agreement between the parties either as to each other's control costs and damage costs, or as to the extent to which pollution of the common resource is to be abated, that there is necessarily a determinate solution. In the upstream/downstream case, B buys the right to its own "clean" environment from A, but there is an area of indeterminacy as to what B should pay A, unless there is agreement between A and B as to A's marginal control cost curve and B's marginal damage cost curve. And, just as in the case discussed above, assumption of agreement between the parties assumes the whole problem away. Even assumption of agreement on guidelines for estimating cost and damage functions removes the right, inherent in the "polluter pays principle", for each country to make use of its own set of preferences in estimating costs and damages. Hence an approach to transfrontier pollution in terms of the purchase of rights to "clean" environment does not remove the problem of deciding how to proceed where different countries have different preferences and where no agreement on common standards is reached.

THE SITUATION WHERE THERE IS NO AGREED PROCEDURE FOR SETTLING TRANSFRONTIER POLLUTION QUESTIONS

Let us next consider what happens if we rely only upon domestic decision of the amount of pollution abatement to be undertaken and upon domestic application f the "polluter pays principle" by each country. We are then in the situation analysed by Muraro, in his paper included in Problems of transfrontier pollution, where cost-sharing between polluter and victim will take place under some circumstances, but where the outcome in any particular case is indeterminate except in relation to the bargaining situation.

However, we are not in practice in the jungle situation postulated in Muraro's paper, because we already have some agreed international principles and a body of international law which are directly relevant to the problem. Bramsen's paper in Problems of transfrontier pollution shows by reference to the relevant cases that international law indubitably requires that states should act in certain specific ways as "good neighbours". He concludes that "harmful transnational pollution is unlawful under international law".

Bramsen draws attention to 1971 UN resolutions on Development and Environment. More recently, Principles 21-25 of the Declaration of the UN Conference on the Human Environment, agreed at Stockholm in June 1972, state that :

Principle 21. States have, in accordance with the Charter of the United Nations and the principles of international law, the sovereign right to exploit their own resources pursuant to their own environmental policies, and the responsibility to ensure that activities within their jurisdiction or control do not cause damage to the environment of other States or of areas beyond the limits of national jurisdiction.

Principle 22. States shall co-operate to develop further the international law regarding liability and compensation for the victims of pollution and other environmental damage caused by activities within the jurisdiction or control of such States to areas beyond their jurisdiction.

Principle 23. Without prejudice to such criteria as may be agreed upon by the international community, or to standards which will have to be determined nationally, it will be essential in all cases to consider the systems of values prevailing in each country, and the extent of the applicability of standards which are valid for the most advanced countries but which may be inappropriate and of unwarranted social cost for the developing countries.

Principle 24. International matters concerning the protection and improvement of the environment should be handled in a co-operative spirit by all countries, big or small, on an equal footing. Co-operation through multilateral or bilateral arrangements or other appropriate means is essential to effectively control, prevent, reduce and eliminate adverse environmental effects resulting from activities conducted in all spheres, in such a way that due account is taken of the sovereignty and interests of all States.

Principle 25. States shall ensure that international organisations play a co-ordinated, efficient and dynamic role for the protection and improvement of the environment.

As far as OECD itself is concerned, there is Article 2(c) of the OECD Convention (1960), under which:

"... the Members agree that they will, both individually and jointly; (c) pursue policies designed... to avoid developments which might endanger their economies or those of other countries."

Member countries are, by the preamble:

"determined to pursue /OECD/ purposes in a manner consistent with their obligations in other international organisations or institutions in which they participate or under agreements to which they are a party."

However, the body of international law and the various UN principles do not help us directly with procedures. They imply subscription to an international approach to the control of transfrontier pollution, but they are insufficiently determinate to tell us how to proceed in the situation where the parties cannot agree on what amount of pollution abatement is appropriate for each and on how the costs should be shared. They do tell us that some solutions are unacceptable. Thus the problem which faces us is to formulate a positive principle or set of principles which will operate consistently with the results of case studies of international law, which will be more specific than the general UN principles and which will allow the formulation of corresponding procedural rules.

15

A POSSIBLE SET OF PRINCIPLES

In Muraro's addendum to his paper in Problems of transfrontier pollution, an extremely interesting procedural rule has been suggested: "don't do to others what you don't do to yourself". This is closely related to the body of international law and to the principles set out (see page 15). But it is incomplete as a procedural rule because it provides no guidance in the situation where one country decides it does not want to abate pollution at all, and pollutes its neighbours' rivers or airspace. The neighbours can of course pay the country to abate its pollution - but it is not the procedural rule which leads to this conclusion. Rather it is the failure of the rule to provide guidance, leading us back to the situation (see page 14), with a cost-sharing solution where the victim pays. Moreover, Muraro's rule enforces high standards of pollution abatement in some cases where there is no presumption that the victim would have abated the pollution had he caused it himself. It is not clear that such a result is just when the polluting country is considered. Hence it appears necessary to search for other principles and procedural rules.

There may be many appropriate principles and procedural rules which are compatible with the UN principles and the body of international law, concerning the way in which a just community of nations should regulate affairs concerning transfrontier pollution between its members. The essential requirements for any such principles are that they should both define the extent of the rights of each party, and that they should indicate the nature of the just course of action in particular cases. The principles should be such as to lead to a feasible set of procedures.

Additionally, the principles should lead to solutions which do not conflict with such guidance as can be deduced concerning the optimum amount or optimum range of abatement. The appropriate test is to examine whether, where there is a determinate optimum, it is reached by the application of the principles. Where there is no determinate optimum, there will be a range, discovered by taking in turn the values of each of the countries concerned, within which the optimum must lie. The point reached by the application of the principles should be within this range in so far as it can be ascertained. In general, there must be no conflict between the consequences of applying the principles, and the consequences which would follow from the international application of the "polluter pays principle", if these can be approximately inferred.

For discussion, a pair of principles is suggested. They do not provide sufficient guidance in cases where the polluted resource is the high seas, since no state has the high seas as part of its own territory. This case requires separate discussion and is not considered in this paper. But for all other types of transfrontier pollution, we suggest first a weaker version of what Muraro has suggested as a procedural rule:

> "no country may pollute another country's resources or a common resource to a greater extent than that which would be allowed by that one of the countries concerned which has the lowest standards of pollution abatement, if the resource were part of its territory."

Once this principle is satisfied, a second principle is applied:

> "no action which benefits a better off country is just unless it can also be shown not to damage the worse off."

Consider the operation of these two principles in an upstream/downstream transfrontier pollution case where B is downstream from A. There are four cases:

i) A is richer than B and sets higher standards for pollution abatement than B.

ii) A is richer than B but with lower pollution abatement standards.

iii) A is poorer than B but with higher pollution abatement standards.

iv) A is poorer than B and with lower pollution abatement standards.

In practice, cases (i) and (iv) are likely to occur most often but there is no reason to think that the others will not be found occasionally. In all cases, we assume that countries apply the "polluter pays principle" domestically to achieve what they regard as their own optimum level of pollution.

Consider case (i). The first principle tells us that A is acting unjustly if it does not abate pollution of B at least as much as B would have done. But it is not required to abate pollution further than this; the second principle requires no more from A.

Consider case (ii). The first principle requires A to make sure that it does not pollute B more than it would do if B were part of A. But the second principle requires more of A; it tells us that A is acting unjustly if A does not abate pollution of B at least as much as B would have done were A part of B, for otherwise rich A benefits from inflicting damage upon B which B would not have inflicted upon itself.

Consider case (iii). The first principle is satisfied provided A abates pollution of B to B's low standards; the second principle is also satisfied by this.

In case (iv), the first principle requires A to abate pollution of B only to its own lower standards and the second principle is also satisfied by this. It follows that if B wants A to reduce pollution of B further, it must make some sort of payment to A before it can expect A to do so, otherwise A will be left worse off by the measures which it takes to reduce pollution in rich B.

These results can be related to the results described in page 13, where countries agree to accept each other's marginal control cost and marginal damage cost functions. In page 13, where B has lower pollution abatement standards than A, because B's assessment of damage enters into the internationally agreed optimum, A abates pollution as much as B would have done, just as in cases (i) and (iii) above. Where B has higher pollution abatement standards than A, the discussion in page 13 indicates that A should abate pollution according to its own marginal control cost curve. In case (iv), where B is richer than A, the presumption is that the marginal costs of control in A soon equal the marginal damage costs; as B wants further abatement in A, it must pay A to undertake it, since the marginal control costs of B so doing are less than the marginal control costs of B undertaking the additional abatement in B, and less than the marginal damage costs. Hence in case (iv) the two principles leads to a result similar to that in page 13.

In case (ii), where the upstream country is both richer than the downstream country and has lower pollution abatement standards for itself, the discussion in page 13 indicates that A should abate pollution according to the internationally agreed marginal damage cost function. Hence it will have to abate pollution of B more than it would have done were B part of A, since B is assumed to value the damage in B more highly than A would, and A has accepted this valuation. We assume that the value assigned to a given marginal expenditure in A is lower than B, since A is richer. Hence the question of B paying A for some of its abatement expenditure will not arise. For case (ii) the two principles solution is again similar to the result in page 13.

Now consider the common resource case. As in page 12, if agreement is reached by the countries concerned about the extent to which pollution of the common resource should be abated, then the "polluter pays principle" applies to each country severally to abate its own emissions so that this standard is attained. Where a common resource is shared by two or more countries of similar income levels, such a solution may well be possible. We do not need recourse to the principles; the lowest standard is the common standard and all countries concerned regard this as acceptable.

But especially where countries have different income levels, agreement on the extent to which pollution of a common resource should be abated may be impossible to reach, except by adopting the lowest standard found among the countries concerned, and other countries will find this unacceptably low. Adoption of the lowest standard satisfies the first principle, but what does the second principle imply in such a case? It implies that the richer countries must pay in one way or another if they want pollution of a common resource to be abated further. They may either pay by abating their own pollution further or they may pay the poor countries to undertake further abatement.

17

Compare this solution with that in page 13. It is important to bear in mind that in the two principles case, there is no determinate international optimum, only a range set by the differing valuations of the countries concerned. In the two principles solution, the richer countries pay for the additional abatement of pollution beyond the level which the poorer countries will accept and pay for themselves. If it is assumed that in general the marginal costs of control for a given abatement of pollution are higher in the poorer country (see page 13), then a solution which imposes the additional cost of abatement upon the richer countries among the polluters will be an appropriate one if there is to be movement away from the bottom end of what can be regarded as the range within which an international optimum might lie.

The discussion above shows and it is important to emphasize this point, that the main difference between the approach to a solution outlined in page 13, and the use of the two principles, lies in the fact that the former method requires that there be a determinate international optimum, whilst the use of the two principles allows this optimum to remain indeterminate and prescribes a procedure for moving towards the range of positions within which any optimum must be presumed to lie. It allows this movement towards and within the optimal range to take place without explicit international agreement on the shape of the marginal damage functions and marginal cost functions of all the countries concerned.

Nonetheless, even with the two principles there remain formidable problems of interpretation. In some circumstances, it is clear that countries would have an incentive to pretend that their own pollution abatement standards in the regions concerned were lower than was actually the case. Machinery to prevent such deception would clearly be required. But to set up and administer such machinery and to reach agreement as to what are the standards which a particular country is actually applying in the relevant regions is a very much easier task than to decide what each country, if rational, ought to do to abate its own pollution.

It may be noted in passing that the two principles could be extended beyond questions of transfrontier pollution to suggest appropriate procedures for the DAC to adopt when considering the environmental aspects of aid to less developed countries. Their implication would appear to be that donor countries should not require LDCs to pursue more stringent environmental policies than the LDCs would themselves choose. If the donor countries wish the LDCs to abate pollution further than this, they would have to give aid, which would be in principle additional to the aid which would otherwise have been given, in order to meet the desired environmental standards. It appears that the DAC is actively concerned about this problem; discussion of the relevance of the two principles approach to it could readily be undertaken.

SUMMARY AND CONCLUSIONS

It is argued that the equation of marginal control costs with marginal damage costs and the subsequent application of the "polluter pays principle" is inadequate to solve the problems of transfrontier pollution because in general there is either no determinate optimum for each country which can be agreed internationally or the right of individual countries to use their own sets of values in determining the amount of pollution abatement which they wish to undertake must be abrogated. (See pages 12-14)

International law and various UN principles require countries to behave as "good neighbours" in certain senses where transfrontier pollution questions are concerned; the problem is to formulate corresponding positive principles which provide general guidance in the different types of transfrontier pollution cases. (pages 14-16)

Two principles to be applied sequentially are suggested for discussion:

i) no country may pollute another country's resources or a common resource to a greater extent than that which would be allowed by that one of the countries concerned which requires the lowest standards of pollution abatement, if the resources were part of its own territory.

ii) no action which benefits a better off country is just unless it can also be shown not to damage the worse off. (page 16)

18

The adoption of the two principles would usually lead to solutions within the optimal range described in a world where there is perfect knowledge but differing valuations of costs and damages, and indeed to the optimum point where this is determinate. However, the two principles can be applied with much less knowledge and a lesser degree of international agreement than is required even to delineate this optimal range, let alone what is required to arrive at a determinate optimum solution. (pages 16-18)

There are formidable problems in the practical application even of the two principles, but they are less than would be involved in agreeing marginal control cost and damage functions internationally. (page 15)

The two principles probably have relevance to the question of the attention which DAC should give to environmental questions when considering aid to LDCs. (page 18).

EQUITY AND TRANSFRONTIER POLLUTION

by

Franco Romani

University of Siena

Italy

The purpose of this paper is to suggest, on the basis of economic analysis, equitable principles for sharing the cost of one-way transfrontier pollution between the polluting and polluted countries.

A typical answer which an orthodox economist (1) might give when faced with this problem would be roughly as follows :

1. One should strive to achieve the optimum level of pollution, i. e. the level at which the marginal damage cost from pollution is equal to the marginal cost of abating it.

2. Different methods could be used to reach this optimum level, the main ones being

    a) a tax on polluters,

    b) a "bribe" from the pollutees to the polluters, and

    c) the establishment of an artificial market for pollution rights.

These methods while equivalent as regards their efficiency, differ from one another in their distributional impact, i. e. with some methods the polluters (or the pollutees) will be better off than with other methods. However, how to decide which party should be better off (i. e. how to define rules on how the various costs due to pollution should be shared) is a matter of equity on which the economist as such has not very much to say.

Nevertheless it is my contention in this paper that the economist who studies transfrontier pollution can go a little further than this. More precisely, I intend to show that the Polluter-Pays Principle is one of the best from the point of view of equity.

The following are the possible principles for sharing pollution costs (2):

   i) the polluter should bear the costs of reaching the desired level of pollu- tion and should be responsible for the residual damage (I shall call this the Civil Liability Principle, or C. L. P.);

  ii) the pollutee should bear the abatement costs and the residual damage (I shall call this the Victim-Pays Principle, or V. P. P.);

 iii) the polluter should pay the abatement costs incurred in reaching the desired level of pollution, while the residual damage should be borne by the pollutee (the Polluter-Pays Principle, or P. P. P.);

  iv) the abatement costs and the residual damage should be borne equally by the polluter and the victim (I shall call this the Equally Shared Respon- sibility Principle, or E. S. R. P.).

The problem now is how to rank these principles, i. e. to establish which one is better from the point of view of equity. For purposes of transfrontier pol- lution I would propose to define an equitable or "just" principle for cost sharing by saying that

"A principle is just if it is the principle which the parties concerned would choose themselves if they had no vested interests."

If this definition is not too unpalatable (3), we shall make some experiments in logic, in which we shall try to see which of the above principles two parties

---

1) See for instance Beckerman's paper in The Polluter Pays Principle, OECD, 1975.

2) Of course one can envisage practical solutions consisting of a combination of these principles. See for instance report at pages 87-114.

3) If the reader is perplexed, he might read J. Rawls, Justice as Fairness, in "Philosophical Review" 1958, for a succinct philosophical defence of this prin- ciple. The point is more fully dealt with in J. Rawls, Theory of Justice, Cambridge, Mass. 1971.

having no vested interests would choose. In the case of transfrontier pollution, the best way to remove "vested interests" from the argument is to assume that, when contemplating the possibility of a certain amount of pollution neither of the parties concerned knows whether he will be the polluter or the pollutee. For the sake of simplicity or exposition, however, it is more convenient to make the stronger assumption that they both have an equal chance of being the polluter or the pollutee (1). If one accepts the value judgement (2) which I have made, it will be possible to employ the analytical tools of economic theory to obtain precise results in the question of equity.

Let us start our analysis on the basis of these premises. Suppose that we have two parties, Tizio and Caio, who are considering the possibility that there may in future be a certain amount of pollution, with an equal probability of either of them being the polluter or the victim, and that they want to decide which principle they should adopt.

Let us consider Figure 1 which is the standard graph showing pollution on the horizontal axis and costs and damages on the vertical axis (where the curve C represents the marginal abatement cost and the curve D the marginal damage from pollution).

Figure 1

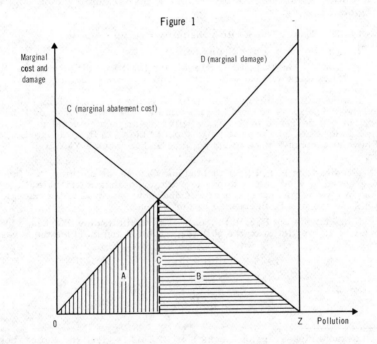

1) Some people consider that the two assumptions are the same, but I would like to point out that my conclusions are also valid when using the weaker assumption.

2) These value judgements would coincide with the value judgements for the Pareto optimum in the case where the two parties we are considering were to reach a lasting agreement on the principles to adopt. In fact, if their "treaty" were to cover a period till "kingdom come", it would become a factual assumption (and not a logical device) to argue that the parties would not know what was going to happen. In this case we suggest as a "good" criterion the one on which there will be unanimity. But taking unanimity as a "good" criterion is merely another way of expressing Pareto optimality (on this point see A. K. Sen, Collective Choice and Social Welfare. Edinburgh and London, 1970 chapter 2).

If one adopts the Victim-Pays Principle and if Caio (and exactly the same applies to Tizio) is the polluter, he will have nothing to pay, while, if he is the victim, he will suffer damage equal to the shaded triangle C (the sum of the triangles A and B).

If one adopts the Civil Liability Principle, the consequences will be just the opposite. If Caio is the polluter, he will have to bear a burden equal to the shaded triangle C, but if he is the victim, he will have nothing to pay.

If one adopts the Polluter-Pays Principle and Caio is the polluter, he will pay the abatement costs (shaded triangle B), while if he is the victim, he will have to bear the residual damage (shaded triangle A).

If the Equally Shared Responsibility Principle were adopted, Caio would always be sure to suffer a loss equal to half the abatement costs and residual damage (one half of the triangle C).

Which of these criteria will be chosen?

In order to explain the behaviour of people when faced with uncertainty, economists have proposed the "expected utility" hypothesis. If we accept this hypothesis (i.e. that people want to maximise their expected utility) as plausible and if, moreover, we assume risk aversion (1), the ranking of the possible principles will be the following:

1. The Equally Shared Responsibility Principle.
2. The Polluter-Pays Principle.
3. The Victim-Pays Principle and the Civil Liability Principle (these two principles will rank equally). This can easily be proved.

If when pollution occurs Caio (and the same applies to Tizio) has an income Y and the V.P.P., is adopted, this income will remain unchanged if he is the polluter, but will decrease by A + B (he will have an income equal to Y-A-B) if he is the pollutee. Just the opposite will happen if the C.L.P. is adopted, i.e. Caio will have an income Y if he is the pollutee and an income Y-A-B if he is the polluter.

If instead the P.P.P. is adopted and Caio is the polluter, his income will be decreased by what he has to pay to abate the pollution (triangle B), while if he is the pollutee he will have to bear residual damage equal to the triangle A, so that his income will be decreased by that amount.

If one adopts the E.S.R.P., he will have his income decreased by 1/2 C, whether he is the victim or the polluter. Table 1 summarises these results.

---

1) To assume risk-aversion means that, if Caio is given the choice between the probability of an income whose average level is $\overline{P}$ and the certainty of an income $\overline{P}$, he will prefer the latter. "This assumption may reasonably be taken to hold for most of the significant affairs of life for a majority of people" (K. Arrow, Essays in the Theory of Risk-Bearing, Amsterdam, 1970, page 200).
"The proposition that risk aversion is the prevalent phenomenon has been defended from personal introspection, from a consideration of the St. Petersburg paradox... and from its success... in giving a qualitative explanation of otherwise puzzling examples of economic behaviour. The most obvious is insurance, which hardly needs elaboration. Common stocks with limited liability to the stockholder find a market because of risk-aversion. The cost plus and other forms of risk-sharing contracts are again explicable only on the same hypothesis... Finally and very importantly in the workings of the economic system the holding of money depends in part at least on the motive of avoiding risks". (K. Arrow, op. cit., pages 90-91).

Table 1

EXPECTED INCOME OF CAIO UNDER DIFFERENT PRINCIPLES

	V. P. P.	C. L. P.	P. P. P.	E. S. R. P.
If Caio pollutes	Y	Y-A-B	Y-B	$Y - \frac{1}{2}(A + B)$
If Caio is the pollutee	Y-A-B	Y	Y-A	$Y - \frac{1}{2}(A + B)$

Now let us consider Figure 2 where on the horizontal axis we have income and on the vertical axis the Von Neuman-Morgenstern utility index. The curve U represents the total utility to Caio of various levels of income and the curvature of U is such that marginal utility decreases (this depends on the risk aversion assumption).

Now if the V. P. P. is adopted, Caio will have an equal probability $(\frac{1}{2})$ of receiving either an income o d or an income o a, i. e. he will have a probable income o e $(\frac{1}{2} o a + \frac{1}{2} o d)$, whose expected utility will be o f.

Figure 2

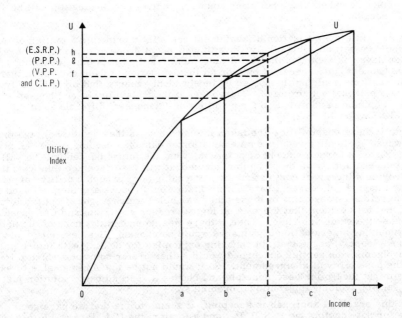

If the C. L. P. is adopted, Caio will have the same probable income o e and the same expected utility o f.

If on the other hand the P. P. P. is adopted, Caio will have an equal probability $(\frac{1}{2})$ of receiving either an income o b or an income o c. Now, as the segment a b is equal to the segment c d (1), he will always have a probable income equal to o e, but with an expected utility always greater than the one he could have either with the V. P. P. or with the C. L. P.

---

1) In fact ad = cd + bd and ab = ad - bd by substitution and we have ab = cd + bd - bd and therefore ab = cd.

25

Therefore, if Caio is maximising his expected utility, he will always prefer the P.P.P. to both the V.P.P. and the C.L.P. This reasoning applies to Tizio as well, so one can argue that Tizio and Caio, sitting round a table and having no vested interests, will both prefer the P.P.P. to either the V.P.P. or the C.L.P.

However, the E.S.R.P. will always be better, because, if it is adopted, Caio (and Tizio) will in all cases be sure of having an income equal to o e with an expected utility o h, which will always be greater then the utility afforded by the preceding principles. So the ranking of the principles will be the following :

E.S.R.P. $>$ P.P.P. $>$ (V.P.P. = C.L.P.).

These are the results derived from the "pure" theory, but we must not forget the problems of practical implementation. From this point of view we have to note that:

a) the E.S.R.P. and the C.L.P. require that the parties should agree on evaluation both of the damage cost and of the abatement costs;

b) the P.P.P. requires only an evaluation of the abatement costs. There may, however, be cases in which it is not even necessary to have an evaluation of the abatement costs. If, for instance, the abatement is, for technical reasons, to be undertaken by the polluting country (as may often be the case), it will be enough merely to fix a standard for the pollution;

c) the V.P.P. requires no evaluation at all, if the pollution control is carried out in the polluted country. On the other hand, if the abatement is undertaken in the polluting country, control cost evaluation will be necessary.

Now, while the evaluation of abatement costs is in principle a matter of fact which could be settled "objectively" by an independent group of experts, the evaluation of damages is clearly a matter of judgement. In fact it is quite reasonable to envisage "honest" differences of opinion on the evaluation of damages, such that possible conflicts could not be solved by independent experts, but probably only by means of a political compromise. Moreover, if the possibility of an "honest" difference of opinion is admitted, the possibility of "dishonest" differences of opinion should also be considered.

But if these possibilities are taken into account (as they should be), the principles whose application requires an evaluation of damages become mere "empty boxes". Given a certain amount of pollution, what the future burden will be will then not depend simply on the principle adopted; it will also depend on the possibility of providing incorrect cost estimates. Pursuing our "expected utility" approach, we could imagine, of course, that Caio or Tizio would choose a principle based on their subjective assessments of these other variables which might affect the future burden, but it is evident that there is no logical reason why Tizio and Caio should both choose the same principle. Then, if there was no unanimity, none of the principles requiring an evaluation of damages would be chosen. Besides, one could also (more loosely) argue that, in choosing principles for dealing with future pollution problems, the parties concerned would rather prefer solutions which minimised the possibility of disagreement. So it would not be too unreasonable to say that one might forego the best principle and choose a "second best" solution just for the sake of avoiding international problems.

If this practical approach is accepted, it seems to me that we have good reasons for not adopting the E.S.R.P. (and a fortiori the C.L.P., which is a bad principle from a practical point of view, as it requires evaluation of abatement cost and damage cost, while from the author's theoretical point of view based on expected utility, it has the defect of being the worse principle with the V.P.P.). So the "second best choice" will have to fall on the P.P.P. or the V.P.P.

With the P.P.P. we still have some practical problems. The major one is that the possibility of different evaluations of damage will effect the choice of the standard level of pollution, so it would seem prima facie that with the P.P.P. we have the same difficulties as we encountered with the E.S.R.P. However, it seems to me that in the case of the P.P.P. there is a way out from the impasse. Thus it is possible to imagine that the contracting parties might choose beforehand, not only the principle, but the pollution standard as well. For instance, they could agree that the standard to be fixed would be the one which the polluter applied to

26

Figure 3

POLLUTER-PAYS PRINCIPLE
(where Caio puts the lower value on the damage
and the standard chosen is the higher one)

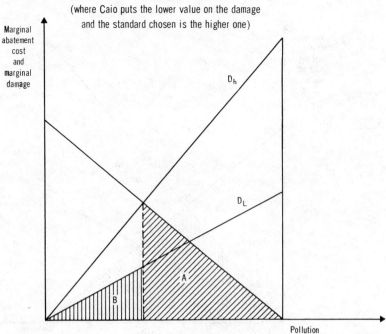

Figure 4

VICTIM-PAYS PRINCIPLE
(where Caio puts the lower value on the damage)

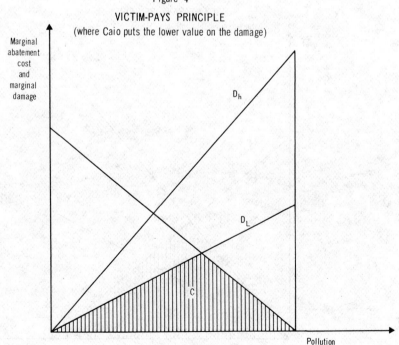

Figure 5

POLLUTER-PAYS PRINCIPLE
(where Caio puts the higher value on the damage
and the standard chosen is the lower)

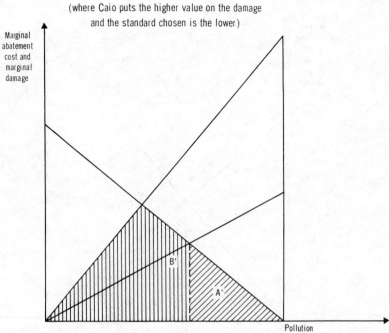

Figure 6

VICTIME-PAYS PRINCIPLE
(where Caio puts the higher value on the damage)

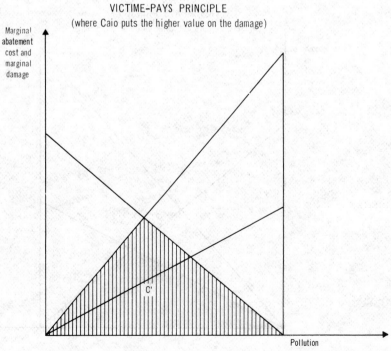

himself, or the one which the pollutee applied to himself, or the higher or the lower of the two (1).

According to our "expected utility" approach, the problem of choosing both the principle and the standard can be analysed in the following way.

Our imaginary party Caio, who has no vested interests, could be either the party who puts the higher value on the damage from pollution, or the party who puts the lower value on it, i.e. he could have either the higher or the lower standard of pollution. However, he cannot know how wide the gap between the two evaluations will be, so that to employ the device of $\frac{1}{2}$ probability for each possibility would not be logical and would also be ethically indefensible. In a situation of this type the way to proceed is to see whether there is a principle and a standard on which there will always be unanimity irrespective of the (subjective) expectations of the parties concerned as to which of them will place the higher or lower value on the damage from pollution.

The possible combinations of cases which Caio may face (and the same also applies to Tizio) are the following:

1. Caio puts the lower value on the damage and the standard of pollution chosen is his own;

2. Caio puts the higher value on the damage and the standard chosen in his own;

3. Caio puts the lower value on the damage and the standard chosen is the higher (i.e. Tizio's);

4. Caio puts the higher value on the damage and the standard chosen is the lower (i.e. Tizio's).

Clearly, if cases (1) and (2) are envisaged, Caio will always prefer the P.P.P. to the V.P.P. for the reasons we have already examined above, but cases (3) and (4) need more careful scrutiny.

We shall first examine case (3). If the P.P.P. is adopted and Caio is going to be the polluter, he will have to pay an amount equal to the shaded triangle A (1), while if he is going to be the pollutee he will suffer damage equal to the triangle B (2). (1) and (2) see Figure 3, which is the standard graph and differs from Figure 1 only because we have drawn two different damage curves, one higher (Dh) and one lower (DL).

On the other hand, if the V.P.P. is adopted, Caio will have nothing to pay if he is the polluter, but he will suffer damage equal to the triangle C (Figure 4) if he is the pollutee. Now it is evident that we cannot say any more a priori and using the "expected utility" hypothesis, regarding which principle is to be preferred to the other (whether, for example, if triangle A is greater than triangle C, the victim-pays principle will be better than the P.P.P.).

Let us now consider case (4). If the P.P.P. is adopted and Caio is going to be the polluter, he will have to pay an amount equal to the triangle A' (Figure 5), while if he is going to be the pollutee he will suffer damage equal to the trapezium B' (Figure 5). If, on the other hand, the V.P.P. is adopted, Caio will have nothing to pay if he is going to be the polluter, but he will suffer damage equal to the triangle C' (Figure 6) if he is the pollutee. It follows from the reasoning in the previous paragraph that the P.P.P. will be preferred to the V.P.P. (2).

---

1) If the abatement is to be undertaken by the polluted country we have also the problem of evaluating the cost of abating pollution to the agreed level, but this problem, as we have already said, could in principle be solved.

2) In fact, since A' + B' = C'; both the P.P.P. and the V.P.P. afford the same probable income, but due to risk aversion the P.P.P. will be preferred, because the probability with the V.P.P. of having nothing to lose will not compensate for the equal probability of suffering a greater loss.

Table 2 summarises these results.

Table 2

	When the lower standard is adopted	When the higher standard is adopted
Caio puts the lower value on the damage	(case 1) P.P.P. > V.P.P.	(case 3) ?
Caio puts the higher value on the damage	(case 4) P.P.P. > V.P.P.	(case 2) P.P.P. > V.P.P.

As we can clearly see from the Table, the unanimity will be reached on the P.P.P. only if the lower standard is adopted. This means that, according to our definition, the P.P.P. combined with the lower standard will be the "just" principle for cost sharing.

I should not be surprised if the word "just" at the end of the last paragraph were considered exaggerated, so before concluding I should like to say a few words in defence of the definition of "justice" which I proposed at the beginning of this paper. I am the first to agree that my definition may not be ethically satisfactory. Why should the behaviour of a person who, having no vested interests, wants to maximize his own benefit be defined as "just"? One can easily argue that the basis of justice is precisely to take into consideration the interests of others, and I personally would agree with this point of view. However, I think that my definition is still-useful and that it helps to establish a "floor" for what is "just". In the case of transfrontier pollution, for example, it defines the minimum which the polluter should be required to pay. To pay less would be "unjust", even if we cannot say that to pay more would be "unjust".

STUDY OF DIFFERENT COST-SHARING FORMULAS
FOR TRANSFRONTIER POLLUTION

Note by the Secretariat

INTRODUCTION

The most difficult problem in connection with transfrontier pollution is no doubt how to determine how countries should share the costs it gives rise to, e. g. the cost of pollution control and the damage cost, when the pollution is at a level considered acceptable by both countries. The problem is dealt with here on theoretical and somewhat abstract lines in order to avoid interfering in any questions now being handled on a bilateral or multilateral basis.

After defining the problem, some criteria will be outlined for choosing a cost-sharing formula. Having recognised that inequality between the powers involved is not the main feature of a description or explanation of international relations, various theories, concepts or principles will be considered, in the light of the criteria selected, which may help in working out a cost-sharing formula. The analysis falls into two parts, depending on whether it aims at resolving an existing situation or at predicting future changes, and mainly concentrates on the case of one-way transfrontier pollution between similar countries. Consideration will then be given to the new factors which are introduced by significant differences in social preferences and levels of development for the choice of the cost-sharing formula.

This report is a sequel to the many discussions on the problem of one-way transfrontier pollution at the Seminar on Transfrontier Pollution organised by the OECD in 1972 (1). In this connection special attention should be given to the reports by R. C. d'Arge, S. C. Kolm, L. J. Locht, G. Muraro, A. Scott and H. Smets on the meaning of equity in connection with transfrontier pollution as understood by economists and to the reports by C. B. Bramsen and R. E. Stein on the legal aspects of transfrontier pollution. This study also takes into account the conclusions of the reports at pages 7-10, 11-19, 87-113.

DEFINING THE PROBLEM

This study deals with the case of two countries which are affected by transfrontier pollution and make a joint attempt in a spirit of co-operation, to solve their common problem by appropriate pollution control measures aimed at abating, but not necessarily at eliminating, the damage to the environment. It assumes in particular that both countries will be prepared to accept a level of pollution higher than the level which would be stipulated in an agreement, provided that compensation is paid for the additional damage, and to abate the pollution below the level stipulated on condition that the extra cost of pollution control is covered. If it is taken for granted that the cost of pollution control and the damage cost are known in the country which bears them in the first instance, both countries will have been able to work out together the best strategy for dealing with the problem of transfrontier pollution and will have agreed upon the advisability of applying it so as to keep pollution down to a certain level in the areas which interest them both (here referred to as the acceptable level, the target level or the level of pollution corresponding to the quality target).

This assertion is based on the assumption that each country is capable of selecting a quality target on a rational basis if it knows the cost borne in the first instance by the other country (cost of waste treatment in the upstream country, damage cost in the downstream country). This choice will be based on the information available, which will not be perfect, but it should not be too remote from the optimum choice which could be made if the knowledge were perfect. In calculating the costs, use may be made of estimates arrived at by a political-type

---

1) See Problems in Transfrontier Pollution, OECD, Paris, 1974.

process, if this process is accepted at national level. Consideration will not, however, be given to cases where the countries have different targets because they act irrationally.

At this stage it still remains for both countries to agree upon possible transfer payments between them (1) i.e. on the formula for sharing the costs due to pollution (cost of bringing pollution down to the target level plus residual damage cost) when the target level is attained. (2)

This study deals only with the sharing of these costs between different countries, and not with the sharing of costs between the national, regional and local authorities within the same country, or between the economic transactors responsible for pollution and the economic transactors affected by it. Most of the findings can also apply to cases of pollution between different regions or provinces in a country with a federal structure.

For purposes of economic analysis we shall assume that sufficient knowledge is available to determine the optimum solution from the standpoint of economic efficiency in the wider sense and that the countries concerned have accepted a formula for sharing additional costs when the level of pollution differs from the optimum level. (3) On these assumptions (4) it would only remain to solve the problem of equity involved in cost-sharing between countries (or determining compensatory transfer payments). It is common knowledge that this problem is not entirely one of classical economic analysis and that it largely depends on value judgements to which the economist can make only a limited contribution. It should further be noted that, whatever the solution adopted and whatever the amounts of the transfer payments between countries, economic efficiency at international level will in principle be safeguarded as long as the polluters and victims of pollution do not benefit directly (or only benefit to a small extent) from the transfer payments and more particularly as long as the Polluter-Pays Principle is applied to every polluter (see in this connection Larry E. Ruff's report in this book).

The cost-sharing formula may vary depending on a number of economic and social parameters. Thus the share borne by a given country may depend on the origin of the pollution and the place where the damage is done; it may vary according to whether the polluted good is appropriable (fresh water) or non-appropriable (air); according to whether one of the countries has started or plans to start economic activities giving rise to pollution, or which are sensitive to it, and are of smaller or greater economic value; according to whether the country has a population with a relatively high social preference for improving the environment and/or a higher standard of living, and to whether it enjoys a very different economic and political status from that of the other country.

The purpose of this study is to examine how the cost-sharing formula may vary according to different parameters and to propose, as far as possible, certain rules consonant with the aims pursued by the various countries and helpful in negotiations between countries dealing with a specific case.

If a transfrontier pollution cost-sharing formula were adopted, countries would at once be able to apply the best strategy and would jointly benefit from a reduction in transfrontier pollution costs (gain from optimisation). On this

---

1) One cannot, for example, set up an international river basin agency for levying charges on polluters in different countries until it is decided how to distribute between these countries the total sums it collects.

2) Problems connected with accidental overstepping of the level are not considered here. They are dealt with in a paper at pages 87-114.

3) In other words, when pollution is one-way, the countries will have come to an agreement on cost-sharing $/C(p)-C(p^*)+D(p)-D(p^*)/$, where $C(p)$ is the cost of pollution control, $D(p)$ the damage cost, p and p^* level of pollution and the optimum level of pollution. However, the countries will still have to allocate the cost $/C(p^*)+D(p^*)/$ (see pages 87-114).

4) In seeking the cost-sharing formula the choice could be made of a level of pollution different from the optimum level. The disadvantage of this approach is that this level has not been defined intrinsically. When the damage cost is not known such a level has to be defined since the optimum level exists no more under this assumption.

assumption, situations would no longer arise where countries failed to reduce pollution to the level corresponding to their quality target pending the conclusion of negotiations, or slowed down the negotiations in the hope of obtaining a more favourable formula thanks to wider national and international support for such principles or slogans as "the polluters shall pay".

CRITERIA GOVERNING THE CHOICE OF A COST-SHARING FORMULA

As transfrontier pollution is a phenomenon of fairly recent economic importance whose overall impact on a country's economy is generally relatively limited it seems legitimate to assume that the formula for allocating costs has not yet been fixed and that countries are still at liberty to choose an appropriate system in line with certain objectives which they are jointly pursuing.

Among the common objectives which may be of some importance in the choice of a cost-sharing formula, the three following may be mentioned here:

a) To avoid encouraging projects which change a country's economic situation and have an impact on the level of transfrontier pollution unless they lead to increased well-being in the countries concerned taken together.

b) To mitigate or eliminate the problems created by transfrontier pollution.

c) To avoid creating undue disparities between areas on either side of a frontier and to endow the frontier areas with a different system from that applied in the interiors of countries.

The first target is of an economic nature and refers to future situations. It implies that a given country should not take decisions which lead to an increase in the total cost of transfrontier pollution, when the change in collective well-being of the two countries resulting from such decisions will be adverse, allowing for the higher total costs due to transfrontier pollution.

The second aim, which is more political than economic, cannot be precisely defined, as it depends on the local and historical context. It implies that a cost-sharing formula should be chosen with due regard to the differing interests of all the countries concerned and that the chosen solution should probably follow a balanced or symmetrical pattern. This aim is based on a series of value judgements, which cannot be overlooked, but on which opinions may differ widely.

The third aim is more social than economic and is calculated to avoid arousing feelings of injustice, unfairness or frustration among the inhabitants of neighbouring areas. It can be fairly easily achieved by arranging lump sum transfers within and between individual countries to counterbalance the financial constraints due to transfrontier pollution and avoid breaks in continuity along regional boundaries. If such transfer payments are difficult to arrange, this aim will be harder to achieve, unless each country has adopted a similar cost-sharing formula for pollution within its own frontiers (e.g., the Polluter-Pays Principle).

In evaluating these three aims it will clearly be necessary to give sufficient weight to the human factors involved in assessing costs due to transfrontier pollution and to take account of the impact of its effects and of the psychological cost of solutions, which are hard to quantify, but are of undeniable political importance, at least in the short term (resistance to change, low mobility of populations, cultural and social aspects, etc.). In this study it will be assumed that the cost-benefit analysis has been carried out so as to give sufficient weight to these factors.

If the cost-sharing formula met these three aims, it might perhaps be considered to be the outcome of a judicious choice. Given the lack of precision in defining the aims, it will probably not always be possible to suggest a specific solution, but an indication of a few general tendencies only may be given.

Furthermore, in making the choice account must be taken of certain constraints or additional aims inherent in the relations between States. To begin with, a number of States are seeking as far as possible solutions compatible with the notion of national sovereignty. This constraint narrows the range of possible solutions and inclines countries not to consider cost-sharing formulas imposed by external authorities, e.g., International Courts of Justice, arbitration boards and advisory boards, a Parliament or a multinational board of directors. However, one should mention the positive role, though with less efficiency, which can be played by a consultation body (1) for the parties concerned. This body could, for instance, take the form of an international federation of national basin agencies for the management of an international river, or a bilateral border commission. It could have representatives of local, regional or national authorities, of various local interests and groups, and experts, but its recommendations would be implemented only with the consent of States concerned.

A further consideration is that some States are sometimes in a position to seek and obtain solutions which mainly satisfy their national interests. Sometimes this objective is counterbalanced by the altruistic aim to reduce inequalities in the distribution of wealth and especially to improve the standard of living in the developing countries. These two aims are obviously unconnected with the problem of transfrontier pollution, and may be dealt with separately in the context of national policies in other fields. Nothing, however, prevents States from negotiating on several issues at once and working out a solution to the problem of transfrontier pollution, while at the same time satisfying other aims by augmenting or reducing certain natural, economic or political inequalities. The same holds good for a country's internal policy, when the Polluter-Pays Principle is sometimes adopted together with a programme of subsidising polluters (transfer payments between polluters and victims of pollution which in fact alter a cost-sharing formula adopted for reasons of regional or employment policy).

In addition, States can negotiate on several matters of common interest at the same time and work out a solution to the problem of transfrontier pollution on lines depending on the solutions of other problems (trade-off). (2)   In this case no strictly uniform cost-sharing formula will be found, even if it existed at the outset of the simultaneous negotiations.

PRESENT-DAY CONTEXT FOR CHOOSING THE COST-SHARING FORMULA

If the basic aim is accepted that countries should try to stop transfrontier pollution from creating problems, a cost-sharing formula cannot be imposed by one country alone and, in view of the constraint of national sovereignty, it can probably not be determined by an external authority either. Consequently, the only possible approach lies in <u>negotiation</u> between two sovereign countries, in the course of which various economic, social and political realities will come to light as many general concepts underlying the international law to which the countries subscribe. The background to the negotiations may be influenced by the "rules of the game" or body of principles which may be proposed as a result of research based on various cases of transfrontier pollution.

While the negotiations are in progress, the difference in economic strength of the different countries cannot be completely overlooked, whether it results from historical, economic or geographical factors and it will be in the interest of the country in the strongest economic situation not to subscribe to any general concepts in respect of cost-sharing, so as to keep the widest possible room for manoeuvre.

If difference in economic strength were the main factor in the negotiations, the country whose economy seemed to be the weakest might feel that the solution it was induced to accept was unfair, and the aim of mitigating the problems caused by transfrontier pollution would not then be attained. Consequently, it might be as well for countries to accept certain cost-sharing rules which were satisfactory for both the polluting and the polluted countries.

In choosing such rules, account will have to be taken of the new-born realisation of countries' responsibilities with regard to transfrontier pollution and of the fact that certain long-standing concepts have been abandoned. Because of the realities of modern society, the time is past when countries accepted treaties placing an absolute ban on pollution, and also when they could claim to have no responsibility for transfrontier pollution. As a whole, the recent trend of

1) On institutions, see A.D. Scott report in "Problems of Environmental Economics", OECD, 1972

2) A "banking" mechanism for the multilateral clearance of payments relating to projects implemented by one country for the sole benefit of another country is described in Annex IV to H. Smets' report in "Problems in Transfrontier Pollution", OECD, 1974.

international law (1) and all the relevant treaties and conventions tend to favour a balance between the rights of the polluting and the polluted countries.

The latest manifestation of international consensus on the environment emerged in the Stockholm Declaration on the Human Environment, when countries agreed to approach international environmental questions in a "co-operative spirit", "on an equal footing" and taking "due account of the sovereignty and interests of all States" (Principle 24). They have the sovereign right to exploit their own resources (Principles 17 and 21), while taking into account the interests of present and future generations (Principles 1 and 2) and ensuring that benefits from the use of non-renewable resources are shared by all mankind (Principle 5).

Excessive pollution must be avoided (Principles 6 and 7), but the quality standards may vary according to the systems of values and the level of development (Principle 23). Principle 7, which is incorporated in the Oslo (1971) and London (1972) Conventions, lays down that:

"States shall take all possible steps to prevent pollution of the seas by substances that are liable to create hazards to human health, to harm living resources and marine life, to damage amenities or to interfere with other legitimate uses of the sea".

The moral obligations of polluting countries are dealt with in Principles 21 and 22:

(21) "States have, in accordance with the Charter of the United Nations and the principles of international law, the sovereign right to exploit their own resources pursuant to their own environmental policies, and the responsibility to ensure that activities within their jurisdiction or control do not cause damage to the environment of other States or of areas beyond the limits of national jurisdiction".

(22) "States shall co-operate to develop further the international law regarding liability and compensation for victims of pollution and other environmental damage caused by activities within the jurisdiction or control of such States to areas beyond their jurisdiction".

As regards relations between the developed and the developing countries (Principles 9, 11 and 12), the developed countries may be required to "transfer substantial quantities of financial and technological assistance", "make resources available", and provide "additional international assistance" for other countries, while ensuring that their "national environmental policies enhance the present or future development potential of developing countries".

The general tenor of the press communiqué issued on the Conference of Ministers of the Environment of the Nine Member Countries of the European Communities (Bonn, October 1972) is similar. Referring to the Stockholm conference, the nine Ministers came to the conclusion that "a global policy for the environment would be possible only on the basis of new and more effective forms of international co-operation, which would take account both of world ecological interrelationships and of interactions in the world economy" ("Une politique globale en matière d'environnement n'est possible que sur la base de nouvelles formes plus efficaces de la coopération internationale qui tiennent compte tant des corrélations écologiques mondiales que des interdépendances de l'économie mondiale"). They considered that action to protect the environment should be inspired by the common principle whereby countries must "ensure", in the spirit of the Declaration on the Human Environment adopted at Stockholm, that activities in one country do not cause degradation of the environment in another country" ("veiller, dans l'esprit de la déclaration sur l'environnement de l'homme adoptée à Stockholm, à ce que les activités dans un pays ne causent pas des dégradations de l'environnement dans un autre pays".) For each type of pollution the level of action (local, regional, national, multinational or Community) should be adapted to the nature of the pollution and to the geographical area to be protected. The nine Ministers considered that as a first stage steps should be taken, for example by consultation, to deal with the environmental aspects of development plans affecting frontier areas.

---

1) See on this point the reports by Bramsen and Stein in Problems of Transfrontier Pollution, OECD, 1974.

These positions, adopted by widely different countries in a world forum, and under the auspices of a Community of developed countries, cannot claim to be entirely new. Indeed, in 1961, the OECD Member countries, recognising the increasing interdependence of their economies, convinced that co-operation would make a vital contribution to peaceful and harmonious relations among the peoples of the world, and believing that the economically more advanced nations should co-operate in assisting the developing countries to the best of their ability, had agreed, both individually and jointly, to avoid developments which might endanger their economies or those of other countries. (OECD Convention).

As transfrontier pollution is liable to affect a country's economy, when the damage caused is significant or the cost of the required pollution control is high, the recent declarations appear to be the natural corollary to a long-standing tendency to encourage the search for a solution to the problems of transfrontier pollution which overrides mere consideration of differences in economic strength and immediate national interests.

CONSIDERATION OF THE CONCEPTS OR PRINCIPLES ON WHICH A COST-SHARING FORMULA MIGHT BE BASED

For analytical purposes the problems of transfrontier pollution will be divided into two categories. In the first, the countries concerned have the same social preferences and would adopt the same quality targets if they were to take over the foreign territory across the frontier affected by transfrontier pollution.

It is easier to analyse the problems of this first category owing to the fact that a smaller number of parameters is involved. In their case it seems reasonable to assume that the marginal utilities of expenditure due to pollution will be the same in both countries.

All other cases fall into the second category of problems, in particular cases where the countries have different social preferences and quality targets and have reached very different levels of development. The solution to the problems in this second category is harder to find and will depend on the solution found for the problems in the first category.

In each category two types of pollution have to be considered, namely one-way (e. g. in rivers) and two-way (e. g. in lakes). As has been shown, a problem of two-way pollution may be broken down into two simultaneous problems of one-way pollution and a solution exists for a problem of two-way pollution as soon as one has been found for the problems of one-way pollution. Consequently the present analysis will deal exclusively with the one-way type of transfrontier pollution, although the problems it raises are without the slightest doubt the hardest to resolve.

The one-way type of pollution, which involves a smaller number of parameters, clearly brings out the divergences between the various "theories" justifying this or that particular cost-sharing formula and calls for a ruling on the applicability of these "theories". On the other hand, cost-sharing formulas for cases of two-way pollution will lack sensitivity to the "theories" adopted by the countries concerned or to the values reached by the parameters which are used in calculating them. Thus, when two countries in similar economic situations share a lake equally between them which is polluted to an equal extent by enterprises with similar characteristics, the waste treatment cost borne by each country will be the same, whether it is paid by the country which pollutes or by the one which benefits from the treatment, but it will be very different if one of the countries is situated upstream of the other on a river. This lack of sensitivity of cost-sharing formulas in cases of two-way pollution, which is a drawback for analytical purposes, is of considerable advantage in finding a practical solution for these problems, as the theoretical divergences between countries have in the end only a slight economic impact.

Use of the various concepts or principles in dealing with problems in the first category

The concepts which may be used in calculating cost-sharing formulas may be classified in two groups, according to whether they allocate all the rights to one country (exclusive concepts), or share out the rights between the countries concerned (non-exclusive concepts).

Exclusive Concepts

Right to discharge wastes

The effect of the territorial concept (or right to discharge wastes) is to authorise the country which first receives an environmental good to pollute it. For example, the first country reached by a coastal sea current or an air stream, or the upstream country along a river, is authorised to discharge pollutants with no regard for the possible consequence to the countries receiving the polluted environment thereafter. Consequently the polluting country should not bear the cost of the pollution control measures which need to be taken for the sole benefit of the polluted countries.

The territorial concept cannot, in fact, be invoked to justify really excessive degrees of pollution, but it does leave polluting countries a great deal of latitude (up to the limit set by the principle of good neighbourliness).

The territorial concept stems from a very wide interpretation given to the right of private ownership. It could be used in the case of river water, since in the absence of any relevant international agreement (1) the upstream country could, in principle, appropriate the water or divert the course of the river, but can be less easily justified in the case of coastal currents of air streams which cannot be diverted.

The territorial concept is also justified by the fact that, owing to their geographical (and therefore historic) location, countries enjoy various special advantages which they are not required to share (natural resources, rainfall, access to the sea, etc.) and that by analogy they are not required to share the dilutive and assimilative capacities of the environment.

The right to a "clean" environment

The effect of the concept based on the quality of the environment is to give to each country, and more particularly to the countries in contact with an environment which is liable to be polluted, the right to an environment no more polluted than the one they would have in the absence of polluting activities, or at least no more polluted than the one enjoyed by the upstream country. It then follows that the downstream country should not bear any cost arising out of transfrontier pollution and that the upstream country should both abate the pollution and compensate a country which is affected by that pollution beyond the "zero effect" level (this concept is analogous to the full polluter liability concept). A limit to this concept is that pollution of the environment below the "zero effect" level cannot be prohibited.

This concept is justified by the facts that the environment was initially clean and that the polluting countries did not negotiate with the polluted countries the acquisition of a right to discharge wastes. Another argument might be based on the notion that air and water are not appropriable goods, but collective goods.

Remarks

The above two concepts, which have the merit of clarity, give exclusive rights to the environment to one of the two countries in cases of one-way transfrontier pollution and are sometimes invoked by countries to justify their negotiating positions. Because of their extreme and paradoxical nature, these concepts maintain an unconciliatory climate. If the negotiations reach a settlement, each country can measure the gap between what has been accepted and what would have resulted from taking one of the two concepts as a starting point.

Furthermore, these concepts ignore the social and economic realities of the areas concerned and do not evolve over time. As they have the disadvantages of

---

1) Such an action would nevertheless be against international law and is only suggested for the purpose of providing an example.

perpetuating a situation, they are unlikely to be recognised by the countries concerned for fear that recognising them might subsequently lead to too unfavourable an economic situation.

It may thus be inferred that the two concepts are not likely to satisfy the aim of mitigating the problems due to transfrontier pollution. However,(1) if countries adopted one of them in their national environmental policies, that concept would assume growing importance in international relations. For example, the principle of fully internalising an externality caused by pollution would amount to at least implicit recognition of the right to a clean environment, while the maintenance of heavily polluted areas would be an implicit recognition of the territorial concept.

The concept of the 'de facto' situation

The acceptance of a de facto situation, i.e. the existence of a right to discharge wastes in places where there is considerable pollution and of a right to a clean environment in places where pollution is negligible, is a synthesis of the two previous concepts which is also found to some extent at national level (industrial zones, recreation zones, etc.). This concept would not easily be accepted by countries if the future development of transfrontier pollution were in practice to result in a single country having to shoulder the financial burden due to pollution. Furthermore, the concept might lead to different solutions being adopted for the problems of pollution in frontier areas and in inland areas, as it perpetuates the rights of the frontier areas to the environment existing at a given moment, whereas the same concept might not be adopted at national level for inland regions.

Non-exclusive concepts

The non-exclusive concepts based on the solidarity principle are very numerous and can be described in a number of ways. In the following paragraphs those concepts will be examined which seem best suited to the problem under study; they include conciliation, non-discrimination, coherency and the sharing of responsibility in cases where social preferences are the same.

The notions of inter-dependence between countries and of solidarity between their peoples, as well as the idea that the environment should be managed collectively for the common benefit, have led to putting special stress on the non-exclusive concepts whereby the right to discharge wastes and the right to the environment are shared between the countries concerned so as to increase the well-being of all countries and individuals.

The notion of ownership or possession of a right to the environment would hereby give way to the notion of using it in a way determined by the resulting benefit to the community. There would be no "tolerable" or "normal" level of pollution as such, but only a level determined by the benefits derived from activities which pollute or are sensitive to pollution.

The concepts based on the solidarity principle are closely connected with ill-defined social, moral and political notions based on certain ideas regarding the equality of the economic benefits from the environment and on the fraternal relations freely entered into between peoples. They tend to allocate costs connected with transfrontier pollution between the different activities which originate the pollution or are affected by it, without examining the "raison d'être" of existing activities. The formulas for allocating costs which could be deduced from the solidarity principle will change over time along with changes in the economic and demographic situation and in people's ideas.

As an illustration of the solidarity principle, one might mention the position taken up recently by an upstream country, viz. "the terms and conditions of financing the works carried out to improve the quality of the river should be governed by the solidarity principle which binds the riparian countries together, i.e. in allocating the costs between these countries, account should be taken of the origin of the pollution and of the benefits which each of them would derive from an improvement in the quality of the water."

---

1) One may observe that the principle of total internalisation (that is recognition of the right to a "clean" environment) is becoming more generally accepted.

Conciliation

Negotiations regarding the cost-sharing formula can be examined from the point of view if its ability to reconcile approaches which are different but not opposite. When two countries are affected by one-way transfrontier pollution, the upstream country will stress the cost of the pollution control measures required to abate the pollution, as well as the lessening of the damage suffered by the downstream country if the pollution is abated. Meanwhile, the downstream country will stress the residual damage cost remaining after the pollution has been abated, as well as the reduced cost of pollution control borne by the upstream country if it fails to bring pollution down to the "zero effect" level.

The upstream country may consider that it should not alone bear the full cost of waste treatment, since the downstream country derives increased well-being from it. The upstream country may try to pay less than the full cost of pollution control and make the downstream country pay more than the residual damage cost.

The downstream country may consider that it should not alone bear the full residual damage cost, since the upstream country derives increased well-being from not having to bring pollution down to the "zero effect" level. It may try to pay less than the full residual damage cost and to make the upstream country pay more than the cost of its pollution control measures.

Between the two positions taken up by the upstream and downstream countries there is only one limiting case in which the two countries will agree. This is when the costs are so allocated that the polluting country bears the cost of pollution control and the polluted country bears the residual damage cost.

If the positions, although close to one another, cannot be brought to coincide at this point, it will be found that the cost-sharing formula will not differ materially from the above formula, but, if they are very different and rigid, the gap may be so wide that the foregoing arguments will do little to help in fixing the cost-sharing formula.

The non-discrimination concept

Much progress would be made towards solving the problem raised by transfrontier pollution if the countries concerned adopted the non-discrimination concept (1) whereby foreign territory affected by transfrontier pollution would be treated not less favourably than territories of the polluting country subject to domestic pollution, assuming similar conditions and the same social preferences in both countries.

By virtue of this concept, one country could not pollute another country beyond the limit which it would itself tolerate in similar circumstances (2) if the

---

1) Various aspects of this criterion were studied by Mr. Kolm and Mr. Muraro, Problems of Transfrontier Pollution, OECD, 1974, and discussed in the report at pages 11-30. The presentation given in the text takes the form of inequalities and serves to find the areas of agreement and disagreement. A somewhat different presentation in the form of equalities could have been made and then the dilemma of areas of agreement and disagreement would be lifted. Tables 2 and 3 are two types of solutions but there exist many others. For practical purposes the non-discrimination concept must be defined in order, precisely, to reach a clear definition of the rights and obligations of the parties.

2) The non-discrimination principle means, among other things, that the pollution of the environment astride a frontier cannot exceed the maximum pollution of that environment in the upstream country and that the quality of that environment cannot be better than the best quality it attains in the downstream country. Given the wide variations in the quality and level of pollution depending on the use made of the environment, these limits are of little value in solving the problems of transfrontier pollution, but better results could be obtained by a confrontation of the policies of countries in which conditions are similar, in so far as it may be possible to define how similar they must be to permit of a comparison.

areas originating and affected by the pollution were both in the polluting country. Moreover, the polluted country should not pay more towards meeting the costs due to the pollution than a polluted part of the polluting country would pay in similar circumstances. Likewise, a polluted country could not demand to be given an environment of a better quality than it would accept in similar circumstances if the polluting area lay within its own frontiers. Furthermore, the polluting country should not pay more towards meeting the costs due to the pollution than a polluting part of the polluted country would pay in similar circumstances.

The non-discrimination concept leads to interesting results with regard to the basis for allocating costs, provided that it is applied separately to both countries in order to determine the areas of convergence and divergence. It sets an upper limit to the costs borne by each country, but without itself indicating what cost-sharing formula to choose.

When each of the two countries has separately adopted the same cost-sharing formula for its own pollution, no uncertainty should remain regarding the formula to be used for sharing the costs of transfrontier pollution. For example, when both countries consider that the polluting area should bear only the cost of pollution control, the upstream country cannot ask the downstream country to contribute to the cost of pollution control, nor can the downstream country ask the upstream country to contribute to the residual damage cost. Consequently, the non-discrimination concept would lead to the adoption at international level of the same principle as is adopted by both countries at national level.

Figure 1

POSITION TAKEN UP BY COUNTRIES IN CASES OF ONE-WAY
TRANSFRONTIER POLLUTION

Costs borne by countries		Upstream country		
		Full internalisation	Partial internalisation	No internalisation
Downstream country	Full internalisation	Waste treatment damage Up ├─────────┤ T Do ──────────────○	Waste treatment damage Up ├─────────────▶▨ Do ▨▨▨▨▨	Waste treatment damage Up○▨▨▨▨▨▨ Do○▨▨▨▨▨▨
	Partial internalisation	Up ├──────▧▧▧ Do ▧▧▧▧	Up ├─────────┤ Do ├─────────┤	Up○▨▨▨▨ Do○▨▨▨▨
	No internalisation	Up ▧▧▧▧▧▧▧▧ Do ▧▧▧▧▧▧▧▧	Up ▧▧▧▧▧▧▧▧▶ Do ▧▧▧▧▧▧▧▧	Up○ T Do○ ├────────┤

▧▧▧ Area of agreement   ▨▨▨ Area of disagreement

┊ line of agreement   T transfer payment   Up. Upstream country
                                                Do. Downstream country

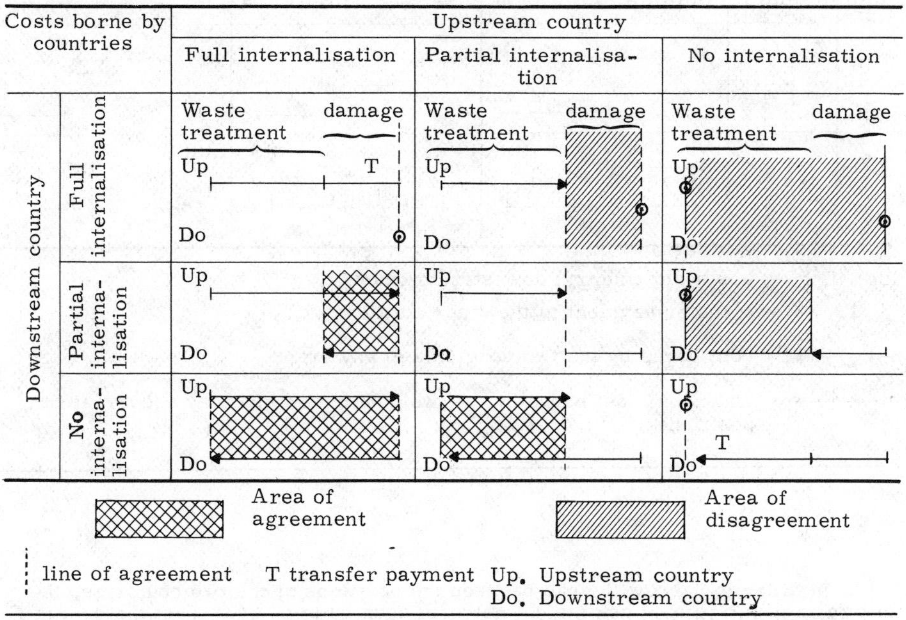

The length of line marked by an arrow indicates the maximum amount which the country shown considers it right to pay, if it applies its own cost-sharing principle and the non-discrimination principle. The point means that the country does not wish to bear any cost. The top left-hand square represents the case where the upstream country is prepared to bear all the costs and the downstrean country no cost. These two positions are mutually compatible and agreement can be reached. The top square in the middle represents the case where the upstream country is prepared to bear the cost of waste treatment and the downstream country no cost; here the disagreement concerns the damage cost. The three principles shown correspond to the principles whereby the polluter must meet all the costs, or the cost of waste treatment only, or no cost, not assumed here is that the cost of waste treatment up to the agreed level is twice the residual damage cost at this agreed level of pollution.

41

If, on the other hand, the upstream country had adopted the principle that the polluted area alone should bear all the costs and the downstream country had adopted the principle that the polluting area alone should bear all the costs, the non-discrimination principle would have left the choice of cost-sharing formula quite open. In the opposite case where the upstream country adopted the principle that the polluting area should bear all the costs and the downstream country adopted the principle that the polluted area should bear all the costs, the formula would not be settled, but the two countries could easily find common ground for agreement. Figure 1 shows the results of applying the non-discrimination principle in the various hypothetical situations.

Figure 2

COST-SHARING FORMULA WHEN COUNTRIES ADOPT MATCHING
POSITIONS

Costs borne by countries		Upstream country		
		Full internalisation	Partial internalisation	No internalisation
Downstream country	Full internalisation	Up ————————T————→ Do ⊙	Up ——————→T⌐⌐⌐⌐▶ Do ⊙	Up ⊙⌐⌐⌐⌐⌐⌐⌐⌐⌐⌐T ⌐▶ Do ⊙
	Partial internalisation	Up ——————→ Do ◄⌐⌐⌐⌐	Up ——————→ Do ◄————	Up ⊙—⌐⌐⌐⌐→ Do ◄————
	No internalisation	Up ⊙ Do◄⌐⌐⌐⌐⌐⌐⌐⌐⌐⌐— T	Up ⊙ Do◄⌐⌐⌐⌐⌐◄—— T	Up ⊙ Do◄————◄—— T

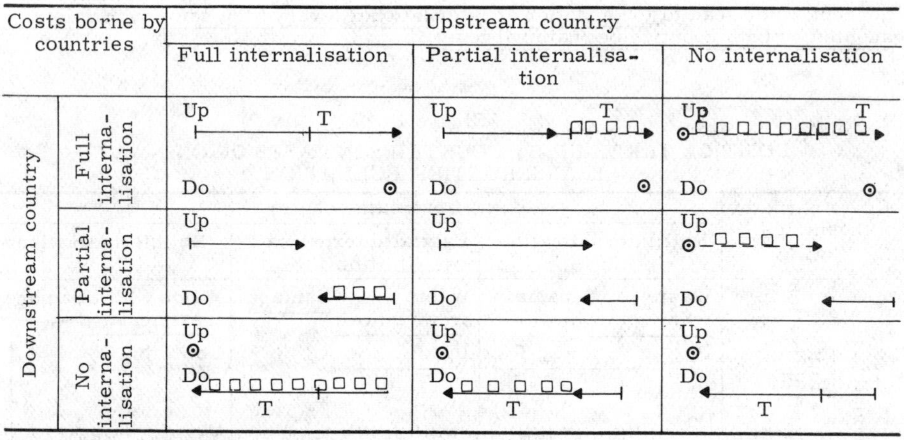

Up. Do. = upstream country; downstream country

T = transfer payment to the other country

————▶ = cost borne by the two countries in any event

⌐⌐⌐⌐▶ = additional cost borne by the two countries when they adopt matching positions

In order to narrow the gap between the positions of the two countries, the upstream country could ask the downstream country to bear at least the cost it would bear if it went by its own principle, while in return the upstream country should bear the remaining cost. This proposition would avoid too pronounced a lack of symmetry between the two cost-sharing formulas (Figure 2).

If the upstream country refused to make any concession in return and managed to keep its contribution down to the amount it would pay by using the most favourable formula, the formulas which would be used in the various possible situations given in Figure 3. There are other possibilities such as resorting to cost-sharing formulas which the upstream (or downstream) countries have adopted for domestic use.

42

Figure 3

COST-SHARING FORMULA WHEN COUNTRIES ADOPT THE APPROACH WHICH FAVOURS THE UPSTREAM COUNTRY

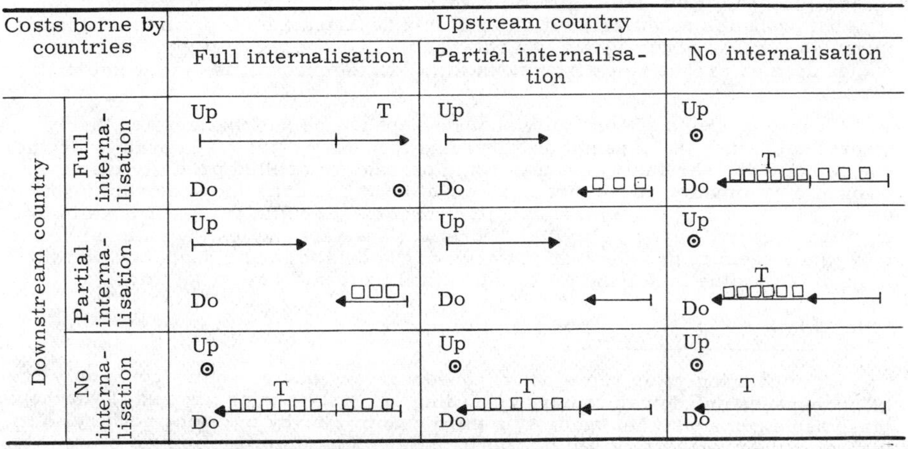

Costs borne by countries	Upstream country		
	Full internalisation	Partial internalisation	No internalisation
Downstream country — Full internalisation	Up ├———————┤——T——→ Do ⊙	Up ├——————→ Do ←◻◻◻┤	Up ⊙ Do ←◻◻◻◻◻◻ ◻ ◻ ◻┤ T
Downstream country — Partial internalisation	Up ├——————→ Do ←◻◻◻┤	Up ├——————→ Do ←————	Up ⊙ Do ←◻◻◻◻◻←——┤ T
Downstream country — No internalisation	Up ⊙ Do ←◻◻◻◻◻◻┤◻◻◻┤ T	Up ⊙ Do ←◻◻◻◻◻├——┤ T	Up ⊙ Do ←┤————————┤ T

Up. Do. = upstream country; downstream country

T = transfer payment to the other country

———→ = cost borne by the two countries in any event

◻◻◻◻◻◻→ = additional cost borne by the downstream country when the proposition which favour the upstream country is adopted

Whatever the solution chosen, it is clear that, in so far as each country adopts the principle that the polluter should at least bear the cost of pollution control, the polluted country should not bear any fraction of the cost of pollution control. The question whether the polluted country should be given a transfer payment in respect of damage suffered could be answered in the affirmative in cases where both countries adopted the principle of full internalisation.

In conclusion, the non-discrimination concept enables countries' obligations in respect of transfrontier pollution to be defined as soon as the countries concerned have adopted domestic policies which are close enough to one another. It does not itself define a cost-sharing formula, but suggests one when each country has adopted a formula for its own use which is deemed equitable. Thus the non-discrimination concept is clearly very useful and there should not be undue difficulty in adopting it, since it means that no country will derive particular benefit from the fact that pollution, by its very nature, crosses frontiers instead of staying at home.

Remarks

Associated with the non-discrimination concept is the concept of geographical continuity, in accordance with which, if both countries have the same cost-sharing formula for their own pollution, the same should be used by analogy for dealing with transfrontier pollution, since it would be the formula used if either of the two countries took over the other one.

There is a kindred concept according to which the formula for allocating costs between two similar countries should be chosen so as not to alter the state of affairs if the frontier should disappear.

These descriptions clearly reveal the judgements of value which underlie the concepts. When the two countries have the same social preferences, the same environmental quality targets and the same rules for allocating costs, the concepts suggest the conclusion that the rules should also be applied to transfrontier pollution, in so far as the two countries have sufficiently similar economies.

When the two countries are not sufficiently similar and differ, for example, with regard to income per head, size of population or total income, it might be right to choose a cost-sharing formula different from that used within each country, but problems might arise if this formula departed too far from the domestic formulas because of too wide a gap between their economic strengths or of demands deemed excessive or irrelevant for reducing inequalities in income distribution.

These concepts of continuity and non-dependence on frontiers are highly theoretical, since the assumption regarding a country being taken over by another one is not taken seriously, but they could perhaps be applied partially inside an economic community whose members were tending to draw more closely together and to be less split by inter-community boundaries. While these concepts can be justified because they simplify the theoretical analysis and lead to rational action, they would seem to provide less powerful arguments for individuals and therefore to be of less value in solving problems raised by transfrontier pollution.

The coherency concept

The coherency (or reciprocity) concept means that a country could only make demands on another country regarding the cost-sharing formula if it accepted these same demands when made by a third country, or by the same country. It mitigates the problems raised by transfrontier pollution by inclining a country to take up the same position when it pollutes as when it suffers pollution. Moreover, it helps to make all countries adopt a formula already accepted by a few countries (snowball effect).

The concept of shared responsibility

Rather than approach the problem of transfrontier pollution from positions which are sometimes far apart, one might choose a global or flat-rate cost-sharing formula without going into causes, effects, legal provisions and responsibilities. This formula might be that the total cost due to pollution should be shared by the countries concerned on the ground that they had a shared or a collective responsibility for creating pollution-initiating or pollution-sensitive activities.

This approach might succeed in eliminating the problems raised by transfrontier pollution and in providing a solution which it would be relatively simple to apply.

There are two types of formulas, depending on whether the total cost is split up on the basis of the incomes of the countries, areas and populations concerned, or on the basis of the actual costs caused by pollution (pollution control and residual damage cost).

First Type

One might, for example, share out the total cost equally between a polluting and a polluted country, or in proportion to the total income or to the income per head or to the size of population in the frontier areas or countries concerned. These cost-sharing formulas would have varying redistribution effects depending on the geographical distribution of income and population and there would always be cases in which they might be deemed to be unfair. Moreover, they would no doubt differ from the formulas used by the two countries for allocating costs between polluters and victims of pollution in different areas within their own frontiers. Despite these criticisms, however, they might be deemed acceptable, since they would distribute "evenly" the burden which transfrontier pollution produces through the interaction between the two countries (1).

Among the various possible cost-sharing formulas, mention should be made of the formula which allocates costs according to the incomes of the areas

---

1) A fuller study of this question will be found Annex III of the report by H. Smets, Problems of Transfrontier Pollution, OECD, 1974.

concerned. These incomes could, if necessary, be calculated from average income per head and the numbers of inhabitants. If the areas were very large because of the widespread interdependence of their economic activities, one might use the figures for countries as a whole. The use of a formula which distributed transfrontier pollution costs in proportion to the incomes of the areas concerned by the pollution would have a levelling effect, since countries could not take a large area as the basis for calculating pollution costs and a small area for calculating pollution costs and a small area for calculating income.

Apart from such more or less altruistic formulas, one might also consider formulas whereby, for example, the total cost to be met by two countries, one of which was richer and more populated than the other, would be divided into equal parts (which would mean a much heavier financial burden per inhabitant in the smaller country).

Second Type

The formulas of the second type are of a more abstract nature, since they depend on calculating each country's share from the costs due to the pollution it causes. For example, the upstream country might bear a fraction of the total cost equal to the proportion which the cost of pollution control bore to the total cost which would arise at the optimum level of pollution. In this case the formula would be to make the upstream country bear the cost of pollution control and the downstream country bear the residual damage cost, when pollution was at a mutually acceptable level. In theory there are an infinite number of possible solutions, but none of them in themselves can be justified as being equitable, so that this approach cannot lead to results of real practical value.

Concepts for use when there is no assessment of damage costs

When the damage cost cannot be estimated, the countries concerned can nevertheless agree to aim at a certain quality target for the frontier area and to follow an effective strategy for attaining it. As the cost of pollution control will be the only quantified factor, the upstream country might try to bear on only a fraction of it and the downstream country might try to avoid bearing any of it, since the latter will already be bearing the residual damage cost.

Although the residual damage cost is assumed to be unknown, one can say that it will be positive and will be below the figure arrived at by taking the absolute value of the marginal cost of pollution control for estimating the marginal damage at that level (this is because the marginal damage cost is usually a non-decreasing function of the level of pollution at all levels down to the target level and because the target level should not be very far from the optimum level). Moreover, while admitting that the damage cost is unknown, the countries concerned will no doubt at least have some idea of the order of magnitude of the damage.

If the residual damage cost turns out to be negligible compared with the cost of waste treatment, it will not be necessary to measure it in order to determine the cost-sharing formula and the polluted country will be able to bear it without trouble. There will then remain the task of sharing out the cost of pollution control between the two countries by using the various concepts described above.

If, on the other hand, the residual damage cost were not deemed to be negligible, it would be more difficult to share out the total cost between the countries concerned, since this cost in unknown.

There are, however, a number of cost-sharing formulas which can still perfectly well be used in this case. They involve making the upstream country pay the cost of pollution control on the ground that they are seen to be fair in dealing with pollution within the upstream country itself, and making the downstream country bear the residual damage cost on the ground that the victims of pollution there bear the residual damage cost caused by that country's own pollution. Another possibility is to let the upstream country bear a fraction of pollution control cost and letting the downstream country pay the difference.

Allocation of costs which result from present and future decisions affecting transfrontier pollution

This analysis has dealt so far with an existing situation involving, for example, pollution produced in an upstream country and a population subjected to it in an downstream country and has led to stating certain principles for choosing a cost-sharing formula which satisfies the aim of mitigating the problems due to transfrontier pollution and the inequalities between neighbouring areas.

In the following paragraphs it is assumed that a formula has been found for the existing situation and a study is made of the new decisions which lead to starting, increasing or reducing a problem of transfrontier pollution.

These decisions will probably be taken by the country which hopes to derive increased well-being from them. If there continues to be a net increase in well-being in both countries, allowing for the rise in the costs due to transfrontier pollution the aim of moving only in the direction of increased total well-being will be satisfied, whatever the formula for allocating the increased transfrontier pollution costs, but nothing warrants the prior assumption that there will necessarily be an increase in total well-being if there is a marked increase in these costs.

To be sure to satisfy the aim of raising the level of well-being it seems necessary to choose between the following two methods: either both countries agree to drop a policy when it is against the general interest, or they accept the cost-sharing formula obtained by applying the causality principle described below.

With the first method, a country whose policy might alter the costs due to transfrontier pollution would have to inform the country affected thereby and both countries would examine together the increases (in well-being and costs due to transfrontier pollution costs). If the result turned out to be negative, the policy would not be pursued.

On this assumption, both countries might consider that they were losing part of their sovereignty. Their decisions would indeed be submitted to a joint body and in addition they would have to instruct this body to calculate impartially any increases in income and in costs due to pollution. The calculation might be difficult because both countries might try to claim high figures for the costs.

The causality principle

With the second method, both countries would accept the causality principle, which may be summed up in the words "he who changes must pay". More exactly, the principle means that a country which alters an activity from which it alone benefits, and thereby alters the externality caused by transfrontier pollution, must alone bear the change in the cost of that externality.

The principle ensures that the addition to external costs is fully internalised and it is quite symmetrical. It is a synthesis of the two opposite concepts whereby the polluters or the beneficiaries alone should pay.

The principle has the merit of not altering the well-being of a country which has no voice in the decision to alter the activities concerned, so that it avoids creating problems. Assuming that both countries take their decisions independently and that the increase in costs due to transfrontier pollution may be significant, it would seem necessary to adopt this principle, if it is desired to achieve the aim of avoiding decisions which would lead to a loss of general well-being as a result of transfrontier pollution.

The following examples make it easier to grasp the implications of the causality principles. If a country pollutes more because it extends its polluting activities, it will bear the additional cost of pollution control and will pay an indemnity to the polluted country so that the latter's well-being may remain unchanged. If a country extends activities which require more waste treatment, it will have to pay the additional cost of the waste treatment and the additional residual damage cost. Consequently the principle will not penalise the polluter if the damage is caused by the polluted country, nor the polluted country if the damage is caused by the polluter.

The causality principle can only be applied when the countries concerned can identify the decisions and investments which require costs related to transfrontier pollution to be revised and when these decisions lead to increased well-being in the country where they are taken. If a cheaper method of treating wastes is invented or a pollutant is found to be more dangerous than expected, the principle cannot be applied, although these developments lead to changes in total transfrontier pollution costs. On the other hand, the principle covers the changes due to the side-effects on the environment of water management works (barrages and canalisation, deepening and changing the course of a river). A fuller list of examples will be found in Table 1.

The causality principle avoids creating disparities on either side of a frontier, since it applies equally to all decisions which alter the cost of transfrontier pollution. It applies to areas, but not necessarily to economic transactors located there and are governed by their national rules and regulations (licensing systems, administrative regulations, payment of charges, polluter-pays principle, etc.). An upstream country which authorised the establishment of polluting industries causing increased pollution downstream should bear the additional damage cost, just as all parts of a country share the cost of any domestic pollution. In other words, the causality principle prevents an area from having to bear a cost arising from a decision which it does not take and from which it does not benefit. If the upstream country did not make its polluting industries bear the residual damage cost, the result might be that decentralised decision-makers and various polluters would launch activities detrimental to the community. By adopting the causality principle one protects downstream areas from the consequences of any such policy - irrational in the long-term - adopted by the upstream country, since these areas would be paid compensation by the upstream community.

However, an upstream country which is late in being industrialised will object that it is obliged to bear a residual damage cost which it would not have had to bear if industrialisation had taken place sooner. It may propose to bear only the additional damage to the optimum level at that moment. The downstream country would then suffer increasing damage and the authorities in the upstream country would be able to set up industries while reducing overall well-being. Accordingly, in the theoretical model under consideration this proposition will not give complete satisfaction, but it may be perfectly justifiable in practice if the additional damage cost in the downstream country is approximately equal to the additional well-being which the downstream country would have derived from the establishment of industries in the upstream country, if transfrontier pollution had remained constant.

The way the causality principle works will greatly depend on an agreement on the initial conditions or on a "normal" level of pollution. One possible method is to take the existing level, but it has its limitations (see also page 50). It might, for example, be held to go too far if a downstream country decided to classify all its hitherto unpolluted frontier areas as low-pollution areas so as to protect itself against any further pollution from the upstream country.

Although adoption of the causality principle will avoid undesirable developments, it will not suffice to ensure the maximum economic development of an area, because decisions taken by a country which appear to be normal and to have no immediate effect on transfrontier pollution may turn out to be costly for the community when new possibilities develop. If, for example, a downstream area can be made into a recreation area and an upstream area into an industrial area, when these two choices are mutually incompatible, the first country to change the use of its land (e.g. the downstream country which develops tourism) will not have to pay any costs connected with a transfrontier pollution which does not exist, and this will prevent the second country from using its land in the new way because of the transfrontier pollution costs which would result (e.g. from industrialisation in the upstream country). If it were more in the general interest to industrialise the upstream country, than to develop tourism in the downstream country, maximum economic development would not be achieved. This example shows that the causality principle is a necessary but not a sufficient condition and that it is in the interest of contiguous countries to consult together regarding long-term plans for developing their frontier areas so as not to have to make choices which would be unsatisfactory in the long term.

The causality principle which governs a series of successive decisions should be generalised or adapted for use in making simultaneous decisions. A possible choice would be to make the countries concerned share between them the additional cost due to transfrontier pollution in proportion to the additional cost which would have resulted from the decision taken by the one country, if the other country had not simultaneously taken another decision. As a rule it is found that the additional cost due to simultaneous decisions is higher than the sum of the additional costs

Table 1

EXAMPLES OF VARIATIONS IN COSTS DUE TO ONE-WAY TRANSFRONTIER POLLUTION BY A RIVER

Case — Origin of change in optimum level of pollution	Effect on well-being if there was no transfrontier pollution ΔB		Variation in total cost due to t.f.p. $(+ \text{ or } -)$ ΔT	Country financing the variation ΔT		Comments
	Up-stream country	Down-stream country		Up-stream country	Down-stream country	
1 — Increase in exposed population in down-stream country	0	+	+	0	+	
2 — Increase in number of activities sensitive to pollution in downstream country	0	+	+	0	+	
3 — Increase in damage when new activities more sensitive to pollution start up in downstream country	0	+	+	0	+	
4 — Reduction in damage because processes less sensitive to pollution are introduced in downstream country	0	+ or -	-	0	-	
5 — Reduction in damage because works are carried out in downstream country for reasons other than environmental protection	0	+ or -	-	0	-	Downstream country alone finances these works and benefits from them
6 — Increase in damage because of secular or unforeseeable climatic or hydrological changes	0	0	+	?	?	In this case agreement must be reached on a cost-sharing formula (e.g., based on relative population sizes or gross outputs from own raw

48

No.	Description						Comments
	...downstream country thanks to its better knowledge of the diseconomies (objective or subjective)	0	0	+	?	?	The downstream country will enjoy increased well-being, because it can follow a more rational policy based on a better estimate of the damage.
8	Increase in number of polluting activities in upstream country (including pollution by consumers)	0	0	+	+	0	
9	Increase in quantity of pollutants discharged as a result of altering the nature of polluting activities in upstream country or of foreseeable accidents	+	0	+	+	0	
10	Increase in pollution upstream due to discharge of pollutants produced outside the upstream area or country	?	0	+	+	0	The upstream area or country may demand payment of a levy on waste discharges before authorising them
11	Increase in damage following a reduction in discharges by upstream country	+	0	+	+	0	
12	Reduction in damage following completion of works carried out in upstream country for reasons other than environmental protection	+ or -	0	-	-	0	Upstream country alone finances and benefits from the works
13	Reduction in cost of pollution control thanks to discoveries due to research financed by upstream country	-	0	-	-	0	Research work will only be encouraged if the country doing it benefits from it.
14	Reduction in cost of pollution control thanks to discoveries due to research financed by downstream country	0	-	-	0	-	Another possibility would be to finance it jointly, if it is successful, on the basis of the cost-sharing formula
15	Reduction in cost of pollution control thanks to discoveries due to research financed by third countries	0	0	-	?	?	In this case a cost-sharing formula must be agreed upon (e. g., based on relative sizes of populations concerned)

occasioned by successive decisions. In some special cases the results achieved in this way will not tend towards the optimum and will be followed by other changes designed to restore a situation more in line with the general interest or the special interests of the countries concerned, (1) so that it appears necessary for the latter to concert their development plans for their frontier areas, even when they adopt the causality principle. (2)

When a change in economic activity or in the use of resources benefits both countries, although it was decided by one country only, the causality principle ought to be better defined. In this case the country which makes the decision should in fact only bear a fraction of the additional transfrontier pollution costs equal to the proportion which the increase in well-being accruing to it bears to the increase in total well-being measured as if the pollution costs had not changed. There is no doubt that there will be practical difficulties in calculating this proportion, but the countries concerned should be able to find common ground for agreement, since they will both benefit from the decision in question. (3)

Finally, if countries used the causality principle long enough for their economic situations to become very different from what they were at the beginning, it might be considered that the agreement had resulted in creating an unfair situation, especially when all the costs were borne by a single country. Consequently the causality principle, although sound as an economic principle, can only be applied in practice to changes which, taken together, do not create a fundamentally different situation from the one on which the agreement was based.

Disagreement on regional planning and choice of economic activities in frontier areas

When countries accept the cost estimates and a cost-sharing formula, they may still find no common ground for agreement if they query the choice of economic activities in the areas affected. The downstream country may consider it wrong to locate a polluting activity next to a protected zone, while the upstream country will perhaps not agree to bear a higher cost, if the downstream country has a satisfactory rainfall and seeks to increase its agricultural output by

---

1) In this connection see the study by H. Smets (Annex III) published in "Problems in Transfrontier Pollution", OECD, 1974.

2) In this context, one should recall the paragraph transmitted by the Stockholm Conference (1972) to the U.N. General Assembly: "Relevant information must be supplied by states on activities or developments within their jurisdiction or under their control whenever they believe, or have reason to believe, that such information is needed to avoid the risk of significant adverse effects on the environment in areas beyond their national jurisdiction". Such paragraph was not adopted in Stockholm because of a disagreement between two States of Latin America although there was general agreement for the setting up of such bilateral notification and consultation procedure.

3) It is difficult to apply the causality principle fairly to past developments, because it involves an internalisation of costs which had not been foreseen when the relevant decisions were taken (principle of non-retroactive effect of laws and contracts). Apart from this consideration, it would usually be difficult to describe, exactly and in the right order, the various decisions taken and to calculate their impact on transfrontier pollution costs from the time when these costs first arose. However, there is nothing to prevent one from fixing the starting date for applying the principle some time in the recent past when a particular event occurred.

If one makes the rather bold assumption that every country's (or area's) present-day income is in proportion to the increase in income it has derived from changes it has made, and also that this increase in income is in proportion to the corresponding increase in pollution costs, the causality principle would make a country choose a cost-sharing formula whereby it would bear a fraction of total transfrontier pollution costs equal to the proportion which its income (or the area's income) bears to the sum of the incomes of the countries (or areas) concerned. In this way we should obtain a cost-sharing formula as suggested in page 44 (shared responsibility).

50

developing irrigation. This problem will not impede future decisions on regional planning if both countries adopt the causality principle, but it may delay reaching agreement on a cost-sharing formula for the existing situation. In practice it is reflected in a desire by one country to calculate the initial cost-sharing formula on the basis of a previously existing situation which would be more favourable to it and then to apply the causality principle.

When the situation in the frontier areas seems abnormal when compared with it would be in other parts of the countries concerned, it would appear to be difficult to ignore this argument completely. One solution might be to change gradually the regional planning policies and development programmes so as to remove the points of conflict between them. The cost of the changes could be shared between the two countries, which would treat them as provisional and exceptional measures as part of a general agreement on the cost-sharing formula.

If the cost of the regional planning changes appeared to be higher than the cost of waste treatment, however, one might conclude that the situation which had been considered abnormal was perhaps not so and that there were strategic reasons for the disagreement on regional planning policies.

Extension of the static and dynamic analysis of the case of countries which have different social preferences regarding the environment. Problems of the second category.

Static analysis

If the countries concerned have different social preferences with regard to pollution, there are two alternatives; either the upstream country sets a lower value than the downstream country on the damage due to pollution, or it sets a higher value. If the upstream country does not consider it necessary to aim at as high a quality target as the downstream country, it will not wish to bear a higher cost than it would bear if the downstream country accepted its lower quality target. The cost-sharing formula should then be calculated as if the downstream country had adopted the upstream country's quality target and would bear alone the additional cost of having the upstream country abate its pollution to a level considered acceptable by the downstream country. If, on the other hand, the upstream country has a higher quality target than the downstream country, the latter may refuse to pay a cost higher than what it would be if the upstream country accepted the downstream country's lower target. The upstream country could lower the pollution level at its own expense, if it thought necessary, or could locate its polluting activities at the frontier, where they would bring the pollution level up to the downstream country's target level. If the downstream country considered that this behaviour was wrong, it could always improve its quality target in the frontier area, thereby eliminating the problem raised by the two different targets and improving the environment.

This solution seems more favourable to the upstream country than to the downstream country, whereas making the upstream country accept heavier obligations because of a collective choice made by the population of the downstream country would raise more difficult problems than those brought about by this solution. However, a better solution might probably be to harmonize quality targets in frontier areas to avoid the difficulties at international level which would arise if different standards were chosen on either side of a frontier.

Dynamic analysis

The dynamic analysis aims at maximising collective well-being and does not bring out the difference between the estimates which the upstream and downstream countries make of the damage done in the downstream country. It takes as the only acceptable estimate of the damage the figure given by the polluted country (if the marginal utility of the costs connected with pollution is the same in both countries) in so far as this figure corresponds with reality (i.e. the figure which the downstream country would give if it bore alone the cost of pollution control).

Extension of static and dynamic analysis to cases of developing countries

In principle, the problem of transfrontier pollution between developed countries and developing countries could be solved in the same way as between countries which have different social preferences, but the same standard of living. However, as was shown in J. Marquand's report in this book, the developed countries might be inclined to choose formulas even more favourable to the developing countries and to bear the main brunt of transfrontier pollution costs, either because they recognise that the pollution was directly or indirectly chargeable to the industrialised countries, or because they wished to increase their aid to the developing countries.

Nevertheless, the choice of a cost-sharing formula is complicated by the fact that the marginal utility of the same expenditure in connection with pollution is not the same in the two types of country and that the problem of efficiency (maximising collective well-being) can no longer be dealt with separately from the problem of equity (1) (sharing out the costs between countries).

---

1) If the well-being functions for two countries affected by one-way transfrontier pollution are

$$W_1 = W_{10} - \alpha\, C(p) - u(1-\beta)\, D(p)$$

$$W_2 = W_{20} - (1-\alpha)\, \frac{C(p)}{u} - \beta\, D(p)$$

when country 1 is the polluting country which carries out pollution control at a cost C, of which it bears the fraction α, and country 2 is the polluted country bearing fraction β of the damage cost, if the two countries have different marginal utilities measured by the parameter u, and if the collective well-being function is $W = W_1 + W_2$, the three well-being functions will be simultaneously at their maximum when $\alpha + \beta = 1$, whatever the value of u. When the pollution level has a value p^* which maximises W, W_1 and W_2, the well-being functions will be:

$$W_1 = W_{10} - \alpha\, (C^X + uD^X)$$

$$W_2 = W_{20} - \frac{1-\alpha}{u}\, (C^X + uD^X)$$

$$W = W_{10} + W_{20} - \left(\alpha + \frac{1-\alpha}{u} \right) (C^X + uD^X)$$

where the asterisk means that pollution is at its optimum level p^*.

The reduction in collective well-being will be independent of the cost-sharing formula (α) if u = 1, and the problem of choosing a formula will be purely a problem of equity between the two countries, with no effect on collective well-being. If u > 1 (country 2 is richer than country 1), the collective well-being will be greater if $\alpha = 0$, i.e. if the richer and polluted country alone bears the total cost. If u < 1 (country 1 is richer than country 2), the collective well-being will be greater if $\alpha = 1$, i.e. if the richer and polluting country bears the total cost $C^* + uD^*$. This shows the difficulties which arise when one assumes the existence of a collective well-being function equal to the sum of the well-being functions of the two countries. On the contrary, if the countries maximise $W_1 + uW_2$ (that is, lay more weight on the well-being of the richer country), the choice of the cost-sharing formula (α) is independent of the difference in living standards related to (u). This ponderation of the well-being functions may be the mathematical expression on the greater weight in joint decision-making a group may carry than another.

CONCLUSIONS

How to define a cost-sharing formula for transfrontier pollution is a difficult question, because it involves countries which are pursuing numerous objectives simultaneously and do not have well established mechanisms for transferring wealth between areas or social groups belonging to different countries. As the question has not only economic implications, it is not surprising if the criteria on which choices are based and the notions of equity which can be used are generally ill-defined and involve many varying value judgments.

It seems possible, however, to suggest a cost-sharing formula, when the countries concerned have similar social preferences and standards of living.

A unifying principle

This study aims at showing that, if countries accept the principles of non-discrimination and solidarity between peoples and adopt sufficiently similar cost-sharing formulas for pollution created within their own frontiers, they could adopt a similar formula for transfrontier pollution. If, on the other hand, they adopt widely differing formulas it does not appear to be possible to indicate exactly what formula might be applicable to transfrontier pollution. As most OECD countries have included the polluter-pays principle in their domestic legislation and as the OECD countries have accepted that principle for dealing with the economic aspects of environmental policy at international level and suscribed to international declarations defining the responsibilities of polluting countries, if seems possible that, for dealing with transfrontier pollution, countries might at least envisage adopting the principle whereby the upstream country alone bears the cost of pollution control (whether it is carried out upstream or downstream) in order to achieve the optimum level of pollution, when this can be defined, or, when it cannot, a target level which is mutually satisfactory.

The following would be the advantages of adopting such a principle:

a) the principle would compromise between the positions taken up by the polluting and polluted countries;

b) the polluting country would not have to acknowledge explicitly the residual damage cost or make a transfer payment to the polluted country as compensation for the residual damage;

c) the general principles of international law would be observed;

d) the importance of frontiers would be reduced, since the same principle would be applied to pollution at home and abroad and it would be easier to avoid distorting the pattern of trade and international investment, since it would not be necessary to create special mechanisms and make transfer payments to even out differences between the rules governing pollution at home and pollution abroad.

This conclusion does not mean, however, that countries would adopt the principle at an early date, because it is arrived at by a theoretical analysis which makes little allowance for the differences in economic strength between countries - a notion which is considered likely to be prejudicial to good international relations - and it overlooks people's resistance to change.

Nevertheless, as transfrontier pollution seems to be a phenomenon of somewhat minor importance, there would be little danger in choosing a fair solution in line with the solidarity principle for international relations, quite apart from other considerations. The adoption of this solution at international level would be the natural sequel to what is being done at national level with the unanimous support of public opinion. The aim would be to establish, as a general objective in allocating transfrontier pollution costs, that there should be a unifying principle, namely that the upstream country should at least bear the cost of pollution control once the countries concerned had adopted this principle in their domestic policies.

Just as at national level the accepted principles are modified and compromises are made to allow for the possible impact of environmental policies on other policies, e.g. regional economic and social policies, so it is unlikely that a unifying principle of this kind would be adopted at international level without reservations being made enabling countries to work out in bilateral negotiations how to apply it in detail, what special transitional measures to take, or even what possible derogations to admit.

He who changes must pay

After solving their current problems, countries could consider taking a position on whether to adopt the causality principle (he who changes must pay) for dealing with future problems. This principle can stop difficulties from arising when transfrontier pollution costs are altered by changes in the economic situation in the areas concerned. It finds justification in economic analysis, favours no country thanks to its symmetry, and generalises the asymmetrical principle contained in the slogan "polluters shall be payers". It can give a lasting quality to any agreement on transfrontier pollution since it lays down in advance what will be the financial liabilities in case changes are made.

Continuing concertation

As has already been shown in this report, the adoption of a uniform cost-sharing formula for solving current problems and of the causality principle for solving future problems will not suffice to solve all the problems raised by transfrontier pollution, so it might be useful to set up in addition a permanent joint planning body to promote the best use of frontier areas and work out the best long-term plans. By arranging a balanced confrontation of the interests of the parties directly concerned, this body could prevent decisions from being taken against the general interest and would enable public opinion to support more effectively measures taken to improve the environment in frontier areas.

Limited length of validity

In view of the changing nature of the concepts on which agreement might be based and because unforeseen circumstances might cause serious disparities between areas on either side of a frontier, or between inland and frontier areas in the same country, it would no doubt be preferable to include a revision clause in any agreement on transfrontier pollution. Nevertheless, it would not be desirable for this revision clause to be able to be invoked too often, lest there should be too much uncertainty regarding the length of the economic life of the investments connected with regional planning.

When countries have different social preferences, the addition of this further point of difference seems to make it much more difficult to find a fair cost-sharing formula. Here the lesser evil would be a solution which gives preference, in calculating the formula, to the position of the country whose quality targets are the less strict, provided that the stricter quality target is achieved. In this case the country with the strictest quality target will alone bear the additional cost of switching from less strict to stricter target, although it would no doubt be preferable to agree to harmonize quality targets in frontier areas.

When countries have not only different social preferences, but also widely differing living standards, it seems to be still more difficult to find a cost-sharing formula, because their environmental policies might be closely connected with their development aid policies. In this case it would appear to be impossible to ask a developing country to bear a greater part of the transfrontier pollution costs than it would bear if the other country concerned had the same level of development.

Part Two

INSTRUMENTS

INSTRUMENTS FOR SOLVING PROBLEMS OF TRANSFRONTIER
POLLUTION

Note by the Secretariat

INTRODUCTION

Whenever a case of transfrontier pollution arises, the countries concerned are usually faced with three problems. First, they can agree on mutual rights and obligations with reference to environmental condition in the area concerned. Secondly, they can decide to give greater weight to the statement of their respective rights and obligations by resorting to implementing instruments. Lastly, they can decide to set up special institutions for strengthening their links because of the increased interdependence resulting from transfrontier pollution and for seeing that recognised rights and obligations are respected as by implementing the agreed instruments.

The purpose of this report is to study different instruments which can be used, on the assumption that the countries have agreed on their rights and obligations but have yet to choose the institutions. The study aims to show the particular advantages and drawbacks of some instruments as opposed to others.

Since transfrontier pollution negotiations appreciably differ from those conducted on a national basis, it appeared necessary to provide in the Annex a brief presentation of the various possible instruments.

DEFINITION OF THE INSTRUMENTS

In this report instruments are defined as the means used to induce the relevant countries to respect the obligations they have accepted in respect of transfrontier pollution. The absence of any formal agreement concerning transfrontier pollution does not however exclude recourse to instruments provided the countries recognise certain principles of good neighbourliness. Instruments which are but means, should not be confused with objectives, which are ends. Thus a standard of quality is not an instrument, nor is a commitment to build a treatment plant or to avoid any pollution. The various types of pressure that a country can bring to bear may be regarded as an instrument only if used to secure compliance with an agreement. If the purpose of pressure is amendment of the agreement or acceptance of a new one, this is not an instrument in the context of this report. International commissions or basin agencies are institutions, not instruments.

Instruments can be accurately defined when, for instance, countries have agreed upon acceptable levels of pollution, the applicable equity concepts, the geographical distribution of pollution control measures and the associated costs for achieving agreed pollution levels, the share of liability when such levels have been exceeded, the sharing of costs when one country proposes to change the agreed level, etc.

Instruments generally create a lasting incentive, one which increases in proportion to the gap between accepted obligations and actual conditions. They are of a preventive and deterrent nature in that they are defined before the occurrence of any undesirable event. They are not coercive, since the countries freely accept them. They sometimes make it easier to readjust to the terms of an agreement when the technological or economic situation changes.

Instruments used between countries are not necessarily like those used by each country in respect of its own polluters and victims. Thus some country might impose a charge on any amount of pollution caused by a polluter in that country while it would pay the other country a pollution charge only for discharges above a certain level. The polluters might also receive from the pollution State a treatment bonus ("bribe") while the polluting state paid a pollution charge to the other country. How a country which pays a charge or bonus allocates it among economic transactors, individuals, basin agencies, local, regional or national authorities, etc. is a subject not here considered.

56

THE NEED FOR INSTRUMENTS

There is no need to resort to any instruments when the countries conclude an agreement which is to be immediately implemented or which can be denounced at any time without major inconvenience to one of the parties. Thus in the case of a polluted international lake equally shared and polluted by the two countries no special instruments need be created, since it is usually in the interest of both to act jointly and in general they stand to lose if they select more limited, non-cooperative solutions.

Nor is recourse to instruments in the event of transfrontier pollution necessary when one of the two countries is not the sole claimant (i. e. polluted) all along the common border and bargaining opportunities are continually found (frontier crossed by similar pollution loads in either direction).

On the other hand, instruments are required whenever one country has short-term obligations and the other long-term obligations. For instance, if a polluted country offers to pay immediately for a treatment plant built in the polluting country or to support the polluting country in some international negotiation (tariff agreement, various concessions, military agreements, etc.) in exchange for a commitment by the polluting country to refrain from polluting an international stream during twenty years, it is necessary for an economic instrument to be adopted in order to induce the polluting country to meet its obligation during 20 years after having received a financial or political compensation at the time of signing the agreement.

The acceptance of effective instruments strengthens the credibility of an agreement and makes it possible to avoid uncertainties which might be harmful to long-term investment and planning in frontier areas. Instruments help to create a climate of mutual trust between the countries, as a real living proof that certain game rules are accepted.

In many cases instruments make it also possible to adjust automatically the terms of an agreement to both countries' satisfaction when a change in socio-economic or technical conditions occurs. By avoiding permanent negotiations the cost associated to these negotiations and the difficulties created by such negotiations can be reduced.

When countries refuse to choose instruments they in fact adopt the principle of continuous negotiation, under which any commitment, however solemn it may be, can be rescinded and renegotiated. In this case the climate of mutual trust can be expected to be less favourable and the countries will be led to choose only short-term solutions which are not always the best from the economic standpoint (e. g. the exchange of immediate advantages, no global negotiations, no setting up of precedents, no mortgaging the future, rescinding an argument at any time, at short notice, etc.).

To sum up, while ad hoc or continuous bilateral negotiation often offers interesting possibilities, recourse to instruments appears to be preferable since the most effective solutions can then more easily be used and disappointment can be avoided when the economic advantage out weighs the moral penalty for breaking the pledge.

DESCRIPTION AND CHARACTERISTICS OF INSTRUMENTS

Instruments may be said to be of three types:

- economic instruments

- legal or semi-economic instruments

- instruments or means of a non-economic nature.

Economic instruments call for monetary transfers between countries and involve direct financial incentives whose use is jointly agreed upon. One typical economic instrument is the charge paid by a polluting country to its polluted neighbour.

Legal or semi-economic instruments are much like their economic counterparts, but must be implemented by a court allocating financial compensation under an impartial decision. A typical semi-economic instrument consists of damages paid in the event of excessive pollution.

Instruments or means of a non-economic nature are the various kinds of pressure which can be brought to bear on the countries but which cannot be directly or clearly expressed in monetary terms (e. g. the desire to honour commitments, the fear of retaliatory measures, the wish to maintain harmonious international relations, etc.).

Instruments are said to be effective when they fulfil their function and yield the result expected in the relevant economic and political context. They are effective and moreover optimal if adjustments promoting the achievement of optimal level of pollution can be made.

Many instruments are ineffective or suboptimal (such as a pollution charge with inadequate rate, a fine too low to act as a deterrent, etc.). Ineffectiveness may also be due to the absence or inadequacy of institutions, to deficient information, slow and costly procedures, or to the scantiness of financial and other resources made available to the bodies responsible for implementing the agreements.

CRITERIA GOVERNING THE CHOICE OF TRANSFRONTIER POLLUTION INSTRUMENTS (1)

Economic criteria

From an economic standpoint the text of a good instrument is maintenance of an optimal pollution level over time through recourse to the least costly strategies of pollution abatement. When the objective (i. e. the pollution ceiling) is defined beforehand, a good instrument must promote use of the best methods of pollution control in whatever area they are applied (organisation of transfers and purchase of treatment facilities).

Furthermore a proper instrument should not entail any unduly high costs of information, administration and negotiation.

Political criteria

From a political standpoint, the more an instrument helps to create smooth international relations and the less it departs from concepts accepted by the countries, the better it will be.

This general criterion for international instruments may be analysed as follows:

a) The instrument should not encourage the countries to take extreme positions in order to protect their interests, resort to threats or call upon any outside authority.

b) It should not prompt the countries to overestimate costs and promote recourse to the pressure of opinion (e. g. press campaigns).

c) It should be balanced and provide advantages (bonuses or bribes) and disadvantages (charges) for each country.

d) It should be a general application or extension at international scale of an instrument which is used or considered acceptable at national scale.

e) It should not require any considerable surrender of sovereignty to some outside body (non-recognition of any outside authority with judicial, coercive or other powers).

---

1) A more detailed study of instruments appears in Annex, and a more thorough analysis of how true cost values are ascertained will be found in report at pages 115-176.

f) It should be capable of being implemented by an institution which is common to the countries concerned and endowed with narrow powers which the countries are prepared to grant.

g) It should not require any third country to inspect, settle or finance certain activities (bilateral problems should be solved bilaterally).

Social criteria

From a social standpoint a good instrument should first enable excessive pollution to be avoided and secondly allow the victims to be thoroughly, reliably and promptly compensated. While the instrument's preventive and deterrent character is more important than its remedial aspect, yet this latter feature must not be neglected.

Such economic, political and social criteria can be used to weigh the relative merits of the instruments. The criterion of equity (the equitable sharing of costs) does not apply here, since an instrument is defined on the basis of previously selected and accepted equity standards. An overview of various possible instruments appears in Annex.

Among them a sizeable number can be dismissed for the following reasons:

a) obvious ineffectiveness: Moral, diplomatic and other types of pressure lead to generally inadequate and sometimes excessive results, since the amount of pressure exerted is difficult to control and the effects are often delayed. Moreover such an instrument may prove largely ineffective if the country where it is applied is insensitive to opinion in other countries or is economically very powerful;

b) worsening of relations between States: Various types of pressure and recourse to international bodies exerting a moral or judicial impact weigh heavily on the international climate;

c) surrender of sovereignty: States reluctantly submit to fines or to any outside coercive authority.

The only instruments which appear to satisfy the various criteria would call for rules setting up a scheme of financial compensation.

The most usual instrument would consist in granting damages to the victims of excessive pollution. The main effect produced by this instrument would be to compensate the victims and incidentally provide an economic incentive inducing the polluter to take adequate pollution control measures. Such an instrument would imply no recognition of liability by a State but merely an international system of transfers set up to restore equity. This does not mean that the government of the polluting State would directly bear the onus consequent upon transfrontier pollution, since financial guarantees could be provided by polluters of that State.

So far this instrument has little been used, yet is accepted by States when really excessive pollution causes undeniable monetary damage.

As matters now stand the instrument is largely ineffective for the following reasons:

a) the amount of damages to be paid is apt to be evaluated in terms of quantifiable and definite monetary damage, while non-monetary damage, loss of amenity and compensation for damage, unimportant for each person but borne by many persons are largely ignored;

b) the procedure for obtaining compensation is lengthy and costly, the burden of proof often being placed on the victims, who often cannot undertake any collective action;

c) when the effects of pollution are temporary or merely inconvenient, the aim is to avoid any renewal of the incident rather than to recover damages;

d) the governments look often for an immediate compromise solution rather than for any later judicial settlement;

e) the polluted States can more easily obtain some concession on a decision later to be taken than immediate compensation for some past event.

This instrument could be made more effective if countries would spell out in advance in an agreement, a set of rules for calculating damages or for determining the price to be paid in case of excessive pollution (agreed upon or contractual compensation). An advantage of this method would be that international problems would be solved in advance and that recourse to damage appraisal by international experts or through arbitration following every excessive pollution incident would largely be avoided also. In this event the payment of damages would become an economic instrument and no longer a legal instrument (like penalties for delays in commercial contracts). However a polluting country could seek to avoid such mechanism in order to gain from the inadequacies of existing compensation mechanisms.

ECONOMIC INSTRUMENTS IN ONE-WAY TRANSFRONTIER POLLUTION

Economic instruments dealing with transfrontier pollution call for financial transfers between States when the level of pollution or treatment is higher or lower than was accepted or agreed upon. They are not used to determine what should be but for rapidly solving problems which arise when the observed situation fails to match a situation of reference. They imply that both parties have agreed on a minimum number of rules. In particular, it will be assumed that each country is able to estimate the cost of damage or pollution control on its own territory and that both countries have settled on a rate of exchange to compare or convert such costs.

In the case of one-way pollution four kinds of situation may be distinguished, according to which cost estimates countries have been able to agree on.

Types		Cost of damage in downstream country	
		Agreement of estimates	No agreement
Cost of pollution control undertaken in upstream country	Agreement on estimates	I	III
	No agreement	II	IV

Type I is an ideal situation very rarely met with in real life. Any efficient economic instrument can in theory be used if politically and socially acceptable. In particular the pollution charge borne by the polluting country above a certain pollution level at the frontier would have the effect of limiting pollution and enable the polluted country to be compensated. In addition the polluted country would help the polluting country if the latter was making a special treatment effort. The pollution charge would be a basin agency charge extended to international scale. It means that the polluting country would choose the optimum pollution level.

Type II is the situation where the cost of damage is well known (similar behaviour of victims on both sides of the frontier) but where the difficulties of determining the cost of pollution control are considerable (many different types of enterprise, regional planning poorly known, etc.). To prevent such difficulties from hindering the implementation of an agreement, the countries might choose such an instrument as the pollution charge, since it is the only one (1) which avoids any exaggerated estimate of pollution control costs. Recourse to any other instrument would necessarily entail a loss of effectiveness and might generate a climate of mistrust. This instrument is being used at national scale for water pollution.

---

1) See report at pages 115-176.

Type III corresponds to the oft-encountered short-term situation where the cost of pollution control is accurately known but where the cost of damage is very differently estimated. The only (1) instrument which obviates any deliberately wrong estimate of damage costs is the treatment charge, whereby the polluted country pays a charge to the polluting country when treatment is intensive and receives a bonus when treatment is slight.

In principle the treatment charge is much like the pollution charge, since in either case the polluting country transfers funds to the other when pollution is heavy, the difference being that the polluted country is in this case the transactor which chooses the optimal pollution level when it is subjected to a treatment charge. The polluting country then sells the service of additional treatment to the polluted country, which is easily able to select an optimal level of treatment upon learning the cost of such a service. This instrument should be quite as acceptable as the pollution charge, at least when the pollution levels under consideration are not too high. The fact that it has so far been little used at national scale should be no deterrent for international purposes, since negotiations are conducted among a few parties which in principle are equally well informed.

Type IV refers to a very commonly observed situation, in which each country concerned by transfrontier pollution begins, before negotiating, by reporting largely different estimates concerning the extent of the pollution and the resulting costs, while it fears that the estimates given by the other country will be exaggerated.

As soon as countries have agreed on an initial situation, it is possible to avoid cost-estimate problems by resorting to the mutual compensation principle (1), that is, by using two charges, one a pollution charge borne by the polluting country and the other a treatment charge borne by the polluted country. This is the only instrument which clears the way to an efficient solution. It is a natural extension of the instruments corresponding to types II and III and should prove quite as acceptable. While it calls for the establishment of a special institution, this would have no extensive power.

Practical considerations

a) Payment of a pollution charge does not imply the recognition of any liability by the State which pays it, but merely a transaction whereby the States adjust differences between real pollution levels and reference levels.

To agree jointly on a price would avoid value judgments and uncertainties. In particular the polluted country would no longer be a victim but become a seller of pollution rights upon collecting the agreed compensation. Such a change of perspective will however be acceptable only if the pollution causes no serious damage to health.

b) The use of economic instead of legal and other instruments would have the advantage of leading to better balanced, more effective solutions.

c) As matters now stand international recourse to a pollution charge would seem possible, since the charge is already being used at national scale in certain countries. Thus two national basin agencies for an international stream might settle on rates to be charged in order to allow for damage caused abroad and might agree if necessary to transfer part of the charges collected under the auspices of an international basin agency federation.

This first step could be used as a stepping stone towards a new system based on the principle of mutual compensation. The chances of achieving economic effectiveness would then be vastly better, since damage costs would no longer be arbitrary or calculated but determined by the victims.

Choice of transfrontier pollution instruments in national contexts

In principle transfrontier pollution instruments governing relations between countries are considered to be independent of national pollution instruments regulating those between the State and polluting (or polluted) parties. This is uncorrect a view, since in all likelihood the countries would try to avoid any undue

---

1) See report at pages 115-176.

disparity between the terms offered to polluting enterprises on either side of the frontier. Moreover it will invariably be easier to transfer funds between States if the transfers are financed by the polluting and polluted parties rather than by the national budget. The conclusion is that problems of transfrontier pollution will be more easily solved if the countries have adopted similar instruments for national pollution and if they adopt these same instruments for transfrontier pollution.

CONCLUSION

Unlike the case for national pollution, in dealing with transfrontier pollution there is a wide variety of instruments to choose from, since only two or more countries are involved. A large number of such instruments can be eliminated as leading to inefficient solutions, as causing exaggerated costs to be reported or as being politically unsatisfactory.

The conclusion reached in this study is that, unless costs are perfectly known and are accepted by the two countries, one instrument alone is ideally capable of meeting the various economic and political criteria. This instrument is a natural extension of the legal tradition associated with damages and applies to a new area of international collective goods. While it appears to be capable of utilisation, some slight change in earlier methods of approach would however be required. The instrument in question, according to the case, would be the pollution charge, the treatment charge or the combined charge, i.e. mutual compensation.

Annex

MAIN INSTRUMENTS APPLICABLE TO ONE-WAY TRANSFRONTIER POLLUTION

INTRODUCTION

The present annex contains a description and brief comparative analysis of main instruments which may be applied in the event of one-way transfrontier pollution. For the sake of simplicity it is assumed that the polluting country suffers no damage and is alone able to take pollution control measures (treatment) while the polluted country suffers all the damage and is unable to counter the pollution.

The instruments may be classified in three groups:

Class   I: Proportional instruments

Class   II: Threshold instruments

Class III: Discrete instruments

This classification depends on the cost function associated with each instrument. The cost associated with proportional instruments thus increases or decreases with the pollution level and is zero at one level only (called exemption level). Threshold instruments have a cost which is zero below or above a certain level called threshold level and which is not zero and increases in absolute value above that level. The cost associated with discrete instruments is zero below a certain level, while it is not zero immediately above that level.

Figure 1 outlines most proportional instruments.

An instrument is used so that a level of pollution will be chosen other than the one probably chosen if the instrument were absent. Instruments may be applied to the polluting country, the polluted country or to both. They create a cost function which is minimised by the polluting country, by the polluted country or both together, or by an agency common to both countries.

Instruments are economic, legal, political or moral, and result in a real or fictitious, monetary or psychological cost for the country to which the instrument is applied. As a general rule economic instruments applicable to transfrontier pollution merely create a transfer between the two countries without affecting the combined wealth of both. This is not true for national pollution, since a government may compel a polluter to pay without necessarily compensating the polluted party.

In the account which follows the three classes of instrument are considered as applying to the polluting country, the polluted country or both together, on the assumption that the well-being of the countries taken together is not affected by the manner of allocating costs due to transfrontier pollution between them and that they seek to minimise the sum of pollution control and damage costs. Moreover, such a sum is assumed to have a single minimum value located between the zero pollution level and zero treatment level (1).

---

1) Mathematically, the sum of pollution control costs C and damage costs D: $C(p) + D(p)$ has a single minimum value at $p^{\pm}$ where

$$0 < p^{\pm} < p_{max}, \quad D(0) = 0, \quad \frac{\partial D}{\partial P} \geqslant 0, \quad C(p_{max}) = 0, \quad \frac{\partial C}{\partial p} \leqslant 0, \quad \frac{\partial^2 D}{\partial p^2} \geqslant 0 .$$

Figure 1

SYSTEMATIC TYPOLOGY OF PROPORTIONAL INSTRUMENTS
(7 effective instruments*)

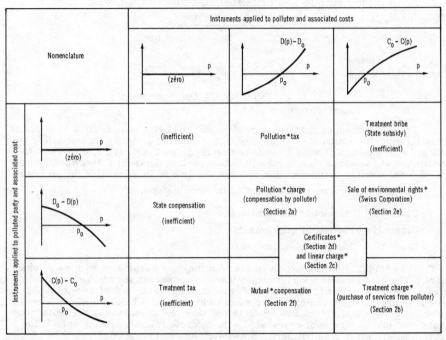

Graphs : In each case, the cost imposed on a country by the instrument is indicated graphically.

$D(p)$: cost of damage
$C(p)$: cost of pollution control or treatment
D_0 : constant term equal to damage cost at reference level p_0
C_0 : constant term equal to treatment cost at reference level p_0

p is the pollution level

NOTE : Instruments involving a pure transfer from the polluter to the polluted party are on the diagonal. Other instruments call for an outside agency body (State, Agency, etc.).

64

CLASS I. PROPORTIONAL INSTRUMENTS (1)

Proportional instruments are charges, taxes and bribes, the purchase and sale of pollution rights (or certificates) or of treatment services. They are represented by a cost increase which becomes zero at a certain pollution or treatment level usually located between zero pollution level and zero treatment level. Proportional instruments as applied to the polluter increase in value with the level of pollution while those applied to the polluted party increase with the level of treatment.

Inasmuch as costs may take on positive and negative values, proportional instruments are symmetrical in character, making them more acceptable to both parties (heavy pollution is penalised but bribes are given for intensive treatment) while they largely lose their punitive character. Generally speaking proportional instruments are purely economic instruments of an incentive and preventive nature. Rather than suggesting any idea of liability, excess or abuse, they are more closely related to a simulated market where both parties engage in transactions according to balanced game rules.

The pollution charge (Figure 2)

The pollution charge is a proportional instrument applied to the polluter, who pays an amount increasing with the level of pollution and receives a bribe which increases with treatment beyond a certain level. The pollution charge may be defined in terms of an "exemption" level where its value is zero and a positive rate (equal to the marginal cost of damage). In complying with this instrument the polluter chooses the optimal pollution level minimising the total cost he bears. The polluted party receives the pollution charge, bears a total cost which does not depend on the level of pollution and hence is not affected by the polluter's choice. At the optimal pollution level the polluted party bears part of the cost of pollution control or is compensated for the cost of residual damage, depending on whether exemption is granted above or below the optimal level (i. e. according to the equity concept adopted). The pollution charge is the instrument corresponding to the "modified civil liability principle" (2). When there is no exemption the polluter pays for any amount discharged (full internalisation of external diseconomies, civil liability) (3).

Note 1) (cont'd)

$$\frac{\partial^2 C}{\partial p^2} \geqslant 0 \text{ and } \frac{\partial^2 (C+D)}{\partial p^2} > 0$$

Under these convexity assumptions, the optimal level is unique and does not depend on the initial bargaining position.

1) A mathematical presentation of the main proportional instruments is given in Annex 2 of report at pages 115-176 where some of their properties are shown. A graphical presentation of these instruments is given in Figures 1-7 of this report.

2) See report at pages 87-114.

3) This particular designation does not mean that it is identical with the associated legal concept, but only that there is some economic relation. See report at pages 87-114.

Figure 2
CHARGE

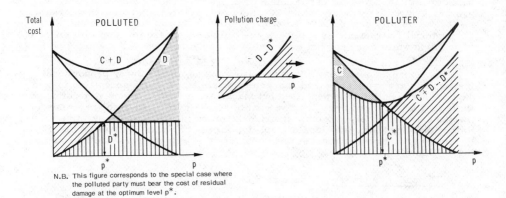

N.B. This figure corresponds to the special case where
the polluted party must bear the cost of residual
damage at the optimum level p*.

The pollution charge can enable an economic optimum to be achieved when
the rate is properly chosen, i.e. when both parties have agreed on the real cost
of damage. In many cases, however, the danger is that the victim of pollution
may estimate the damage cost at a higher level than the polluter will accept. It
can be shown that if the polluted party alone is granted the right to estimate the
marginal damage cost and if exemption approaches the optimal level, the polluted
party will give an estimate which to the best of his knowledge closely approximates
the true value in his own economic and political context (1).

The pollution charge does not require both parties to agree on estimates of
pollution control costs and would seem to be an ideal instrument when the parties
agree on damage costs (2). An advantage is that it needs no adjustment in the event
of pollution control cost changes over time, while an accompanying disadvantage
is that the optimal pollution level is not specified.

As pollution charges are being increasingly used at national scale, they might
also be internationally applied, although so far they have not been. Thus when two
countries share a river basin and each makes use of such an instrument under a
national basin agency, both countries could optimally manage the basin by choosing
charge rates for the national agencies which would take the international character
of the basin into account.

If the countries retain the charges collected by the national agencies from
national polluters and the latter equate the marginal cost of treatment with the rate
of the charge, then the two countries provide for optimal basin management based
on the polluter pays principle (the polluter bears the cost of pollution control measures and the polluted party the cost of residual damage when the pollution level
is the acceptable level). A transfer of the charges should only take place in case
the downstream country is accidentally subjected to excessive pollution or tempo-
rarily enjoys an environmental quality above optimum. (At the accepted level, no
transfer is necessary.)

Should the countries between them agree on an equity principle such that the
polluted party is not expected to bear the cost of any residual damage (full internalisation, civil liability) and if the downstream country alone suffers damage, then
the charges collected by the upstream country from the polluters causing damage in
the other countries should be transferred to the downstream country, regardless of
the pollution level reached.

---

1) The instrument will however only remain efficient if the rate of the charge can
   be altered over time when the cost of damage varies.

2) See report at pages 115-176.

<u>The treatment charge</u> (Figure 3)

The treatment charge (1) is the proportional instrument applied to the polluted party which corresponds to the pollution charge applied to the polluting party. The polluted party pays the polluter a positive charge for treatment above an agreed exemption level and receives a bribe for treatment under this level. The rate of the treatment charge is the absolute value of the marginal cost of treatment. The polluted party bears a cost which is at a minimum at the optimal pollution level while the polluter bears a cost unrelated to the pollution level. In this system the polluter undertakes to control pollution for an agreed price up to a level determined by the polluted country (sale of treatment services). When the treatment exemption is zero, the treatment charge is the instrument corresponding to the victim-pays principle (2) and the polluter is paid for any treatment he undertakes.

Figure 3

TREATMENT CHARGE

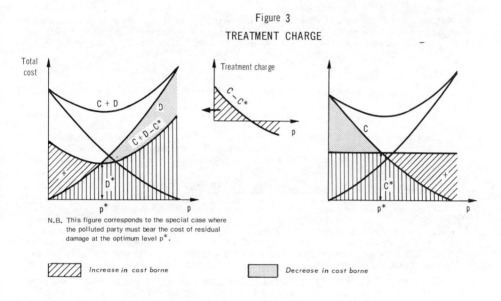

N.B. This figure corresponds to the special case where
the polluted party must bear the cost of residual
damage at the optimum level p^*.

[diagram legend] Increase in cost borne   Decrease in cost borne

The treatment charge may enable an economic optimum to be achieved when the rate is properly chosen, i.e. when the two parties have agreed on the real cost of pollution control measures. It does not call for any agreement on the cost of damage and it will not be in the polluted country's interest to abate pollution below the optimum. It would thus appear to be an ideal instrument when the cost of pollution is accurately known while the cost of damage is not (3). An advantage is that it can be adapted in case the cost of damage changes over time, while a concomitant drawback is that the optimal pollution level is not specified.

The treatment charge seems to be little used in dealing with national pollution. As the instrument requires the victims to express a collective opinion regarding damage and to ascertain an economic optimum, it is hardly surprising that it should not be used when the victims are many, poorly informed and unorganised.

---

1) Treatment charge is not to be confused with treatment bribe (Figure 1). The bribe is paid by an organisation not representing solely the pollutees while the treatment charge is paid by the polluted parties to the polluters.

2) See report at pages 115-176.

3) See reports at pages 87-114 and at pages 115-176.

In the case of transfrontier pollution the victims are represented by a government which is perfectly capable of making rational choices. For this reason the instrument deserves to be regarded as particularly suited for resolving difficulties in the event of disagreement on damage costs. The case of accidentally exceeding the pollution level is dealt with in page 74.

The optimal linear charge (Figure 4)

The optimal linear charge is linearly proportional to the pollution (or treatment) level, whose rate is equivalent to the marginal cost of damage (or to the absolute value of the marginal cost of treatment) at the optimal pollution (treatment) level. By means of the optimal linear charge a transfer between the polluting and polluted parties is arranged in such a way that the net costs borne by the polluters and the victims are constant or minimum at the optimal pollution level.

The optimal linear charge can be analysed as a combination of the pollution charge and treatment charge (1).

The linear charge may be used in place of the pollution charge (or treatment charge) provided the polluting (or polluted) party as a marginal treatment (or damage) cost which increases. An advantage of this instrument is that information is needed for only two quantities :

- the exemption level (equity)
- the constant rate (efficiency)

It will hence be simpler to use than pollution or treatment charges which require more complete information. Conceptually the countries will only have to determine upon a unit price for the purchase or sale of "pollution rights" instead of having to decide whether the polluter is responsible for any additional pollution and the polluted party for any additional treatment required.

The purchase and sale of "certificates" (Figure 5)

The certificate system, first suggested by Dales and later taken up by others, including Scott (2) is based on the creation of certificates entitling the holder to discharge pollutants into the environment. It is assumed here that each holder causes the same amount of damage upon discharging the same amount of polluting matter, that the certificates are allocated among the polluting and polluted countries and that the countries are empowered to buy and sell the certificates.

If the polluting country holds fewer than the number of certificates it needs for emitting quantities approaching the optimal pollution level, it can purchase them from holders in the polluted country, who will sell them at a price which is higher than or equivalent to the cost of additional damage but lower than the cost of additional treatment which is avoided. Under conditions of perfect competition between the holders, the polluting country pays no more than the cost of additional damage and the certificate system is identical with the pollution charge. If competition is imperfect the certificate system more closely resemble the treatment charge.

---

1) The pollution charge applied to the polluter is equivalent to the damage cost $/D(p) - D(p_1)/$ measured from a determined level p_1 and the treatment charge $/C(p_2) - C(p)/$ for the polluter is equivalent when $p > p_2$ in absolute value to the cost of treatment not done. The combination in correct proportion of the two increasing functions $D(p) - D(p_1)$ and $C(p_2) - C(p)$ with second derivatives of opposite sign generates approximately a linear function whose first derivative is a rate for the optimal linear charge equivalent to the marginal damage cost and to the absolute value of the marginal treatment cost at optimal level. When the marginal cost of damage or treatment is constant the optimal linear charge is identical with the pollution or treatment charge.

2) See Problems of Transfrontier Pollution, OECD, 1974.

Figure 4

OPTIMAL LINEAR CHARGE

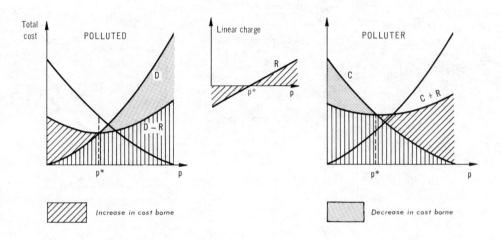

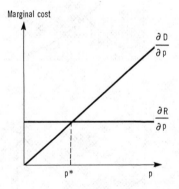

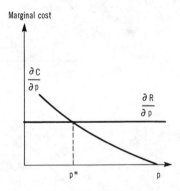

Figure 5

PURCHASE AND SALE OF CERTIFICATES

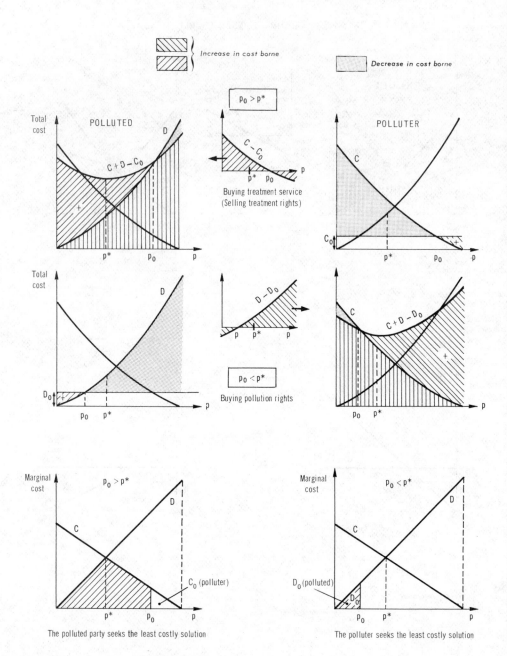

Increase in cost borne

Decrease in cost borne

$p_0 > p^*$

POLLUTED

$C + D - C_0$

D

$+$

p^* p_0 p

$C - C_0$

p^* p_0 p

Buying treatment service
(Selling treatment rights)

POLLUTER

C

C_0

p^* p_0 p

$+$

Total cost

D

p_0 p^* p

D_0

$+$

$D - D_0$

p p^* p

$p_0 < p^*$

Buying pollution rights

C

$C + D - D_0$

$+$

p_0 p^* p

Marginal cost

$p_0 > p^*$

C

D

C_0 (polluter)

p^* p_0 p

The polluted party seeks the least costly solution

Marginal cost

$p_0 < p^*$

C

D

D_0 (polluted)

D_0

p_0 p^* p

The polluter seeks the least costly solution

70

If the polluting country holds more certificates than needed for meeting the optimal pollution level, then it will be able to sell certificates to the polluted country at a price higher than or equivalent to the cost of additional treatment needed and lower than the cost of additional damage avoided. Under conditions of perfect competition between the holders the polluted country only pays the cost of additionally needed treatment and the certificate system is identical with the treatment charge. If competition is imperfect then the system resembles more nearly the pollution charge.

It will therefore be seen that, depending on the conditions of competition and on how the certificates have initially been allocated, the certificate scheme is either similar to the treatment charge or to the pollution charge. An advantage of the certificate system is that it requires no prior knowledge of cost functions, since the certificates need only be allocated for the scheme to set rates. A drawback is that the optimal level is not specified and will not be reached if the certificates are bought and sold on an imperfectly competitive market (particularly in the case of transfrontier pollution between two countries when certificates are held by the two governments).

So far the system appears not to have been used at national scale (1). In addition it does not offer any particular advantage over the principle of mutual compensation in the event of a bilateral monopoly. Use of the certificate system in dealing with transfrontier pollution would probably call for its prior acceptance by countries desiring to solve national pollution problems in an international basin affected by transfrontier pollution.

The sale of environmental rights or the Swiss corporation (Figure 6)

If the countries assign to a joint agency the power to sell the polluting country the right to discharge pollutants for a price equivalent to the cost of the treatment avoided, and the polluted country the right to treatment for a price matching the cost of the damage avoided, then both countries bear constant costs regardless of the pollution level. Assuming that the countries own the joint agency in proportions to be defined, it will be to their advantage to see that profits are maximised by the joint agency, which will hence be induced to choose optimal pollution level. Under this system the joint agency is seen as the owner or holder of all environmental rights and can sell the countries certain advantages drawn from the common asset at a maximum price, so that the two countries do not directly benefit from optimisation. This system corresponds to the "Swiss Corporation" presented by Scott (2) when the cost directly borne by the polluting country is the maximal treatment cost (zero pollution) and the cost directly borne by the polluted country is the maximal damage cost (zero treatment).

So that this instrument can operate the joint agency must know the maximum price it can charge for environmental rights, which means that difficulties will arise in the case of two countries (a bilateral monopoly, between the agency and each country) controlling the agency's operation unwilling to accept any unduly "mercenary" approach by the joint agency's authorities. It can be shown (3) that the system cannot lead to economic efficiency when the countries have failed to agree on cost estimates and that it does not induce them to declare the exact value of the costs they bear in the first instance.

The system of selling environmental rights is based on the simultaneous use of two charges and seems to offer no special advantage in transfrontier pollution when compared to the system of mutual compensation. It has not yet been used at national scale and hence presumably should not be adopted for dealing with international problems.

---

1) With the exception of construction certificates in France where the owner of a land can buy or sell the right of building so many sq.m. of floor per sq.m. of land.

2) There appears to be only one existing case where countries have agreed on the principle of joint ownership of a common resource (Lake Titicaca, between Peru and Bolivia). See Lester, Am. J. Int. Law, 57, No. 4, 841 (1963).

3) See Problems of Environmental Economics, OECD, 1072, p. 263.

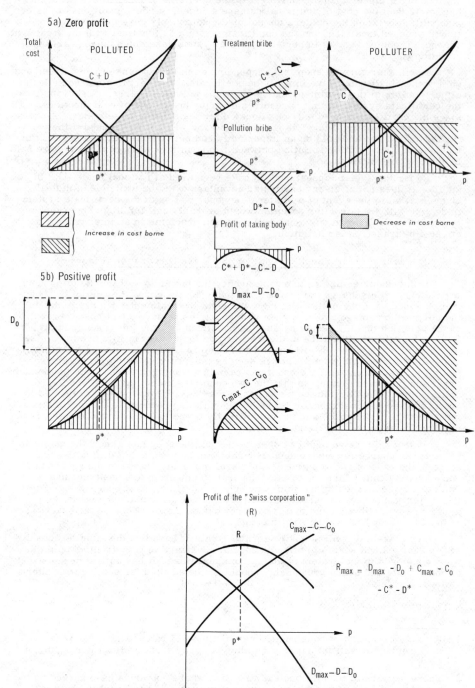

Figure 6
SALE OF ENVIRONMENTAL RIGHTS

5a) Zero profit

5b) Positive profit

The optimal pollution level p^* is selected by the corporation which sells to the polluted party the advantage of a purer resource and to the polluter, the advantage of less treatment.

Mutual compensation (Figure 7)

The system based on the mutual compensation principle resembles both the certificate scheme and the Swiss corporation approach, yet is distinct from either. It is based on the simultaneous use of the pollution charge paid by the polluting country and the treatment charge paid by the polluted country, both collected by the joint agency. In this case the cost borne by each country is minimum at the optimal pollution level and both will require the joint agency to determine such an optimal level by minimising the sum of the charges collected from them. If it is assumed that each country knows the cost initially it bears and pays an amount equivalent to the cost declared by the other country, it can be shown (1) that the system induces each to determine and announce the exact value of the cost it initially bears. Lastly, economic optimality is attained even when the countries fail to agree on cost estimates. The mutual compensation system emerges as the only instrument which can be used for achieving economic optimality when the countries have not agreed on cost estimates.

Figure 7

MUTUAL COMPENSATION PRINCIPLE

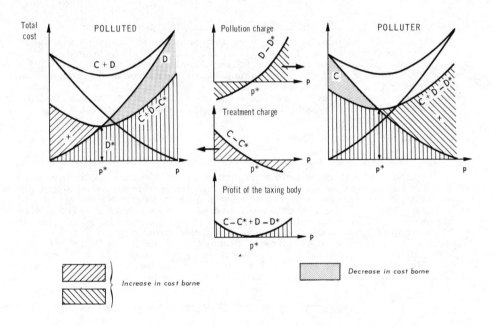

It can be put to use as soon as each country agrees to pay a charge and no recourse is needed to outside experts for "impartially" calculating the "real" cost of damage or of treatment pollution control. The joint agency acts in a purely executive capacity according to established rules and not as a monopolistic owner (as for the Swiss Corporation) trying to sell the economic benefits of a common resource at the highest price.

The mutual compensation scheme has not so far been used but offers such advantages that it is well deserving of attention in the case of transfrontier pollution, where negotiations are conducted between two organised parties representing the collective interests of their constituents.

1) See report at pages 115-176.

Other economic instruments

There are many other instruments which might be used in transfrontier pollution but whose merits have been considered inadequate compared with the schemes described. These in particular consist of instruments coming under the principle of shared responsibility (sharing of the total cost: C + D), the sharing of treatment costs, non-linear instruments in cost functions, instruments related to auctions of certificates, etc. The case where an agency outside the two countries might collect a tax from one country without transferring it to the other could also be analysed.

Nor has the use been analysed of such generally inefficient instruments as those resembling tax credits, reduced interest rates and accelerated depreciation. In the case of two countries one might thus offer the other to bear part of the cost of treatment.

Neither have we considered the redistributive charge, which can be used only when polluters are collectively compelled to avoid exceeding some pollution level (a redistribution device among polluters rather than between polluters and victims).

Lump sum subsidies are not dealt with in this analysis since they correspond to a lump sum transfer between countries and are included in each case where a zero charge level or exemption level is defined. That any third country or organisation should be willing to grant a subsidy seems an unrealistic assumption.

Conclusions with regard to proportional instruments

To conclude, three proportional instruments can be advocated as economically efficient when the countries fail to agree on all cost estimates. When the countries agree on damage costs but not on pollution control costs, then the pollution charge would seem to be the best instrument. When they agree on pollution control costs but not on damage costs, the best instrument would appear to be the treatment charge. And when they do not agree on any cost estimates, the best instrument would consist in using both the pollution charge and the treatment charge (mutual compensation).

Whenever instead the countries do agree on estimates for both costs, many other instruments are possible to use. These may have certain political and social advantages (acceptability) or economic advantages (cost of implementation).

From a practical standpoint no proportional instrument has so far been used internationally and few have been used at national scale. The recourse to proportional instruments will therefore create special problems dependent on resistance to change. The particular properties that some instruments have of enabling economic efficiency to be achieved without calling for an agreement on all costs must doubtless be regarded as the best economic argument promoting the use of such instruments. An additional advantage of proportional instruments is that they require no further international testing of legal concepts such as liability which are extremely awkward to use when damage to public goods, loss of amenity, and variously adverse effects on a great number of people are the outcome.

CLASS II. THRESHOLD INSTRUMENTS

The most typical threshold instruments are compensation payments or penalties imposed when the pollution (or treatment) level exceeds the threshold level. Generally their purpose is to keep the level under the threshold, while the associated cost is zero below the threshold and increases in absolute value above it. Practically speaking threshold instruments are no different from proportional instruments when the level is always above the threshold. They are not symmetrical since the associated cost is constant below it. In certain cases a lump sum transfer may be combined with the threshold instrument so that the net associated cost below the threshold is negative or positive.

In some respects threshold instruments are punitive in character for the party who pays while reparative with respect to the party receiving payment. Above all they are linked to "abnormal", "excessive", "abusive" or "accidental" situations and consist of legal or semi-economic instruments (allocated or agreed compensation) and instruments or means of non-economic nature (various types of pressure).

Threshold instruments may also take the form of deterrents with respect to the polluter without necessarily creating any legal relationship between the polluting and polluted parties if the latter is compensated through a collective compensation fund. Under these conditions the polluter would no longer be "guilty". He would pay no compensation in the event of excessive pollution but would only have to pay a compulsory premium to a collective compensation fund adjusted according to the degree of precaution and the preventives measures taken by the polluter.

The payment of damages for excessive pollution (Figure 8)

Theoretical analysis

Making the polluter pay an amount which increases as a permitted maximum level of pollution is exceeded is a long-established traditional method of inducing the polluter to comply with a standard. The case will here be examined of a polluting country which compensates the polluted country, since it seems hardly realistic to assume that the polluting country would agree to pay a fine or penalty to some outside international agency.

Figure 8

DAMAGES

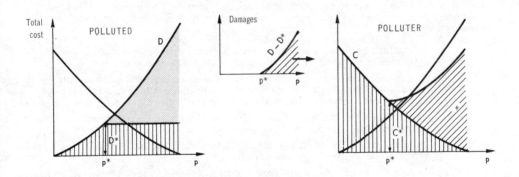

If the polluting country knows that it will probably pay a compensation it will take measures to control pollution such that the absolute value of the marginal cost of treatment equals the rate of compensation. The level of pollution reached will be optimal if the rate of compensation equals the marginal damage cost at such a level and if the threshold level is under or equal to the optimal pollution level. If the rate of compensation exceeds the marginal damage cost at the optimal level and the threshold is equal to the optimal level, then the instrument is equally efficient. If the compensation rate is below marginal damage cost, then the instrument will be inefficient and the level of pollution reached will be above the optimal level.

This instrument can only be considered equitable provided the rate of compensation to the polluted party is equal to the marginal cost of damage. It can only be applied when a threshold level is determined by agreement or judicial decision (1).

---

1) As a rule countries have not defined a maximum tolerable level of pollution by international agreement. There are a few cases, however, where such levels are directly defined (Agreement between Belgium and the Netherlands on the Ghent-Terneuzen Canal, Treaty on Pollution of the Great Lakes). In other circumstances the countries have adopted standards (radio-active pollution in the European Community).

An agreement in principle to grant damages beyond a certain established pollution level does not in itself lead to economic efficiency or equity unless the countries settle on some way of calculating damages meeting the above criteria. The calculation should take account of intangible aspects and of the polluted environment regarded as a public good. To compensate only those victims who complain for the amount of direct damage capable of being evaluated in monetary terms results in a systematic undervaluation of the rate of compensation (no account is taken of victims who fail to complain or who are unaware of the damage they suffer, while succeeding generations are ignored). A better solution in dealing with transfrontier pollution might consist in globally compensating the polluted State, which would be left the responsibility of possibly compensating victims individually itself.

Application to transfrontier pollution

The payment of damages to pollution victims creates difficulties whenever the polluting country consciously tolerates any higher level of transfrontier pollution. Such a hypothesis is not considered, since lack of agreement on objectives rather than implementation of an instrument is involved.

To the extent that the objective is accepted by both countries, then compensation of the victims is a generally acknowledged concept of equity. Yet the payment of damages is a problem difficult to solve on the international scene, unless the "guilty" polluter can be clearly identified and is solvent, the damage is considered to be extensive and the victims are identified.

A current although not yet general trend consists in requesting the polluting country to pay damages to the victims while leaving it to take action against the polluting parties identified as "guilty". Such a trend is justified for the following reasons :

a) The polluting State as a whole derives direct economic benefit from the activities of polluters while the polluted State does not in general.

b) It is the responsibility of the polluting State (either as a general rule or by reason of particular agreements) to see that polluters on its territory do not cause any undue pollution in neighbouring countries.

c) The activities of polluters are usually subject to permission and control by the polluting State.

d) The polluting State must see that polluters do not arrange to claim insolvency in the event of serious accident.

e) As a rule the polluted State cannot determine the causes of excessive pollution without interfering with the operation of government in the polluting State or with the management of the polluting firms.

While the responsibility of the polluting State may appear normal in the case of pollution which is not purely accidental, it might be contested in the case of major episodes concerning which the polluting enterprise and polluting State have already taken many safety precautions for internal reasons.

Actually, however, it has been observed that States appear less inclined to assume financial responsibility for minor pollution casualties than for those reaching disastrous proportions. Underlying this attitude may be the idea that no State can subject another to a new, extensive risk (full internalisation in the case of new activities so as to counteract the added factor of loss of absolute security by the obligation to pay compensation). Little account has moreover been taken of inextensive transfrontier pollution problems because the countries have gradually come to accept increasing amounts of pollution both within and along their frontiers and are not interested in minor disturbances to the neighbouring environment.

In practice it is also noted that countries rarely (1) accept financial responsibility where transfrontier pollution is concerned for reasons connected with the

---

1) Recognition of liability

Most States have explicitly or implicitly recognised their liability under international law for really excessive pollution, but many seem to be desirous of avoiding any formal recognition of financial liability for transfrontier pollution which does not reach disastrous proportions.

notion of national sovereignty and because they are often unwilling to assume greater responsibility for transfrontier than for national pollution. This attitude no longer strictly matches development in international law during the past twenty years, since States have recently formally acknowledged their liability for radioactive pollution and for the effects of space activities conducted on their territory (1).

The difficulty of explicitly recognising State liability can be partly got round by emphasising the payment of compensation to the polluted country by the polluting country on grounds of international solidarity and equity (2) and by avoiding the the more philosophical issue of the polluting State's liability, which now only acts as a transferring agency. By means of this bias the desired result is obtained with regard to the polluted country and the issue of internalising external costs is shifted from the international scene to the national scene (search for the best strategies of pollution control).

Note 1) (cont'd) :

Yet the 1922 Copenhagen Convention between Denmark and Germany established the principle of both civil and penal liability in the event of any damage caused, in particular by discharging waste matter in frontier waters (Art. 33), while the 1960 Treaty between the Netherlands and Germany provides for the payment of compensation (Art. 63) and a special procedure for settling differences.

1) On this score reference may be had to an article by J. M. Kelson: "State Responsibility and the Abnormally Dangerous Activity", Harvard International Law Journal, 13 No. 2, pp. 197 - 214 (1972), in which the writer attempts to show that States should assume strict liability for dangerous industrial activities and proposes three principles:

a) Duty of the State to take measures in order to prevent harm caused by dangerous activities;
b) Duty of the State to notify a neighbouring State of dangerous activies;
c) Strict liability of the State for harm caused abroad.

It should, however, be pointed out that the concept of strict liability is new and differs from traditional law, where the emphasis is on fault and on the moral and financial responsibility of the person committing the fault.

See also L. F. E. Goldie: "Liability for Damage and the Progressive Development of international Law", Int'l Comp. Law Q., 14, pp. 1189-1264 (1965).

2) Compensation cases

There are few cases where countries have demanded or obtained compensation for transfrontier pollution. In particular one should quote the following cases:

- Trail Smelter: payment in favour of the United States after international arbitration for air pollution;

- Japanese fishermen and residents of Rongelap: payment by the United States of compensation after diplomatic negotiations following radioactive pollution;

- Swiss farmers: payment of compensation for air pollution by a factory in German territory;

- Torrey Canyon: compensation of the French and British Governments following an accidental oil pollution.

A more comprehensive study was made by Prof. du Pontavice (Restricted OECD document).

Equality of treatment

One way of partly getting round the obstacle of recognition of the State's liability in international law is to grant foreign victims of pollution the same rights as national victims in a similar case, while authorising the foreign victims to take individual and collective action under the same conditions as nationals in national pollution cases. Provisions granting the same rights to foreign victims are contained in the 1909 Treaty between the United States and Canada and in the Nordic Convention on the protection of the environment (1974).

An advantage of this principle of equal treatment is that the polluting State carries no heavier obligation than in the case of national pollution, but a drawback is that compensation of the foreign victims is linked to the law and jurisprudence of the polluting country (intangible damage is often poorly covered). In addition it is difficult to see a polluting State granting indirect financial advantages to some foreign polluted area, for example in the form of environmental improvements paid for out of the polluting State's own budget, while such action is conceivable for reasons of national solidarity.

An international agreement on equality of treatment might however be a first step in effectively implementing the damages instrument in transfrontier pollution. It would gain in effectiveness in case the national courts treated the polluting parties with greater severity, as present trends seem to indicate. The signature of such an agreement would seem a real possibility especially between neighbouring and similar countries united in alliances or members of international organisations.

The transfer of penalties, taxes and charges

When polluters of the polluting country pay penalties, taxes or charges in the event of excessive pollution, the polluting country would easily be able to arrange compensation for the victims of transfrontier pollution by transferring all or part of the penalty, tax or charge amounts. Transfer might be made under the auspices of an international basin agency or directly between governments and would involve no formal recognition of liability.

Collective compensation for pollution risks

The scheme whereby polluters are compelled to contribute to a collective fund compensating for pollution risks implies that a common agency be set up by polluters with adequate financial responsibility for covering exceptional cases of extensive damage caused by pollution (1). Under this system the victims can be rapidly compensated without their having to determine liability. An advantage is the solidarity and collective responsibility thus created among the polluters, thus releasing the remainder of the community (in particular the public authorities) from the onus of compensating the victims of insolvent or unidentified polluters. Moreover the polluters must bear the consequence of any irregular conduct on their part (non-compliance with laws and regulations by certain polluters) and the State is freed from liability for ineffective surveillance, inaction or neglect in the exercise of its prerogatives with respect to the polluting parties.

In the case of transfrontier pollution it would be for the collective fund rather than the State to compensate the victims upon application by the polluted State acting for the latter.

The system will not create any real incentive for polluters unless the amount paid by each into the fund is adjusted in terms of the risk. Such a risk might be evaluated by members of the polluting industry, who would be held collectively liable or by actuaries which the industry would select. Were the State to be made additionally liable in the event of extreme damage (State guarantee), it would then be induced to take such safety measures as to prevent its having to act as guarantor.

---

1) This system is an institution rather than an instrument. It is described here owing to the difficulty of implementing this instrument when recourse is had to the judicial bodies existing in all countries.

In some ways this scheme (1) resembles compulsory insurance, since compensation due by the guilty party is always paid by the insurance companies. It differs from the traditional concept of liability in that compensation is paid without proof being required that some individual polluter is at fault and even if the excessive amount of pollution is due to events over which the polluters have no control (such as unusual weather conditions). With this method the risks created by their own polluting activities are entirely borne by the polluters. But it is not applicable when the victims help to increase the damage by acting in some manner considered to be inappropriate (shared liability).

The system of compulsory insurance for nuclear installations operators linked to the concept of strict liability of such operators was internationally introduced in 1960 by the OECD to cover cases of excessive radioactive pollution (Convention of Third Party Liability in the Field of Nuclear Energy, Paris). Under this system victims are compensated through an insurance fund up to a maximum amount without its being necessary to prove that the operator of the nuclear installation is at fault. The scheme moreover provides that the operator shall be solely liable, thus inducing him to take any needed measures with a full knowledge of the facts.

Thirteen countries concluded a supplementary convention (Brussels, 1963), which states that governments would be responsible for covering risks in excess of the maximum amount specified in the Paris Convention (for damage amounting to between $15 and $120 million per incident). Were the insurers to default, then governments would themselves have to pay the compensation due by the insurers (Vienna Convention, 1963). The system thus set up at international level is supplemented by many provisions of national scope so as to ensure that victims will be paid compensation through government financial action designed to complement the responsibility of operators.

This method of compensation in the field of nuclear energy is an original concept under international law and might perhaps be used as a model for other pollution of an exceptional character (e.g. accidentally excessive pollution due to toxic chemical substances, to oil spillage during extraction, production or transport activities, or to toxic industrial wastes). This example shows that States can join in devising new instruments to meet the requirements of modern society even when no disastrous incident has yet prompted them to innovate.

More recently three conventions have been signed in Brussels as regards oil pollution by tankers. They institute a system of strict liability with a ceiling and set up a mandatory collective insurance fund (2). Efforts are underway to set up a similar system for sea and shore pollution due to drilling-platform accidents.

Transaction costs

Insofar as damages are not automatically allocated to the victims such an instrument involves fairly high costs of litigation, administration and negotiation as well as considerable delays in some instances (sometimes ten years). Recourse to the instrument will be more frequent and less costly if the following arrangements are made:

a) recourse by the polluted States (or polluted groups) is admitted on behalf of all victims;

b) liability is channelled towards one person (the operator of the polluting installation, the State, etc.);

c) compensation takes place as soon as the pollution exceeds a predetermined threshold (few exceptions restricting the right to compensation);

1) In some parts of the United States, a similar scheme is already implemented for car accidents (no fault insurance) and results in saving litigation costs.

2) The private agreements TOVALOP and CRISTAL between tank owners and oil companies offer polluters the advantage of a mutual insurance to cover pollution risks below $40 million. Such agreements are the proof that a collective fund for compensation is a realistic proposal which can be implemented. In 1974, a private agreement OPOL offers a similar system for platforms in the North Sea.

d) damage costs are estimated in terms of an agreed procedure or scale;

e) a body exists which is competent to settle disputes promptly and conclusively if they cannot be quickly resolved by negotiation (1).

Purchase of treatment services (Figure 9)

In the case of transfrontier pollution the polluted country might choose to pay the polluting country compensation for additional treatment beyond an accepted level. This instrument is the <u>purchase of treatment service</u>. The associated cost is zero at normal treatment level and positive beyond. The marginal cost of the instrument is the marginal cost of treatment. Hence the polluter bears a constant cost for treatment above the normal level and the polluted party chooses the additional level of treatment by minimising the cost he bears in terms of the compensation paid for additional treatment.

Figure 9

PURCHASE OF ADDITIONAL TREATMENT SERVICES

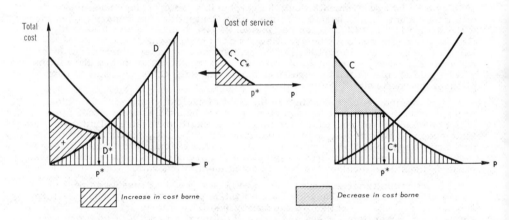

In order to be efficient, this instrument must be used in conjunction with another instrument such as a fine (Figure 12) or penalty for excessive pollution (Figure 10). This latter system (2) named "modified victim-pays principle" has the advantage that it is in line with the principles of international law when pollution is extensive and that it induces the polluted party to ascertain the cost of damage when pollution is slight, whereupon the intangible aspects become very important. It is therefore of particular interest whenever additional damage cost (above the normal pollution level) and the cost of pollution control can be assessed without too much difficulty by the countries.

The instrument is used at national scale when the polluted parties are able to negotiate with the polluters and to grant payment for treatment (purchase of a service) and the polluters are bound not to exceed a certain pollution level. It may be likened to direct negotiations between countries when the polluting country agrees to respect a certain standard. Despite its merits it has not yet been explicitly used in transfrontier pollution.

Instruments or means of a non-economic nature

Instruments or means of a non-economic nature are more or less strong methods of persuasion which may be associated with measures of pressure. They are used with respect to a country which makes "excessive" or "abnormal" claims.

---

1) See B. M. Clagett: Survey of Agreements providing for Third Party Resolution of International Waters Disputes. <u>Am. J. Int. Law</u>, <u>55</u>, No. 3, pp. 645-69, 1961.

2) See report at pages 87-170.

Figure 10
PURCHASE OF TREATMENT SERVICE/DAMAGES

9 a) Zero transfer at the optimum

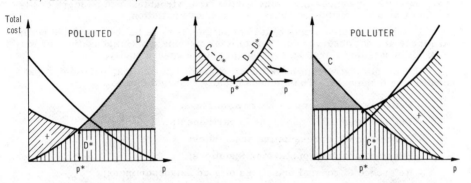

9 b) Non zero transfer at the optimum

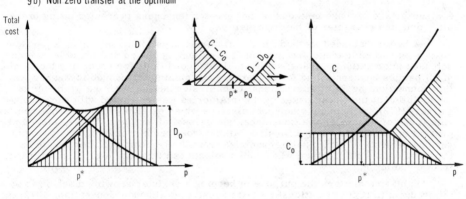

The pressure of opinion

An excessive polluter or a polluted party demanding excessive quality may be subjected to moral pressure from outside, prompting him to adopt a more "normal" attitude so as to preserve an adequate international reputation (or a satisfactory "brand image"). The instrument constituted by such moral pressure may induce countries to align their behaviour with their declared intentions or written commitments. While not highly efficient this instrument plays no uncertain role in transfrontier pollution (excessive or inadequate response of opinion, insensitivity of the country at fault to public opinion, misunderstanding of the facts).

The greater the amount of available information, the better it can be checked, and the more closely countries will have defined their obligations, then the more efficiently the instrument can be applied. Thus a pledge to refrain from discharging toxic substances into some particular environment can be more easily verified than a commitment to avoid its excessive pollution.

Pressure of opinion may cause commitments to be better respected but in case these are not observed does not in itself provide any compensation for victims. It is not, therefore, an instrument able to ensure equity.

Use of this instrument calls for all sorts of information, publicising and conditioning techniques. The methods usually resorted to are:

- Petitions, meetings of concerned citizens;

- Public statements, questions in parliament;

- Resolution of representative assemblies;

- Press campaigns, public manifestations;

- Reference to general principles of good neighbourliness;

- Reference to codes of good behaviour, universal declarations, treaties, etc.

A disadvantage of these methods is that greater tenseness is created owing to the often oriented character of the information.

A better-balanced method for the countries consists in setting up a permanent or ad hoc joint agency responsible for collecting and publishing objective and unbiased information on transfrontier pollution conditions for recording the opinion of all parties concerned and for acquainting the public with needed measures. In this connection the Joint International Commission dealing with the problems of the pollution of the Great Lakes of North America was entrusted with such prerogatives (1972). The many other commissions which exist play a somewhat similar role (Lake of Geneva, Lake Constance, the Swiss-Italian lakes, Rhine, etc.) and prevent to some extent any difficult relations from being created by reviewing the various projects of participating countries from the beginning (1) (through prior consultation and notification) when such projects can increase a problem of transfrontier pollution.

In marine matters the purpose of keeping a register showing discharges of oil products (IMCO conventions) or toxic substances (London Convention, 1972) is to exert moral pressure on the polluters.

In the nuclear area the Commission of the European Communities, which must be notified of any discharge of radioactive effluent, has the right to check on the operation and efficiency of radioactivity monitoring installations, and if obligations are not complied with by the Member States the case may be put before the Court of Justice.

Economic, diplomatic and other types of pressure

Countries (and their nationals) may voice their opposition to the conduct of some other country by resorting to measures of pressure (e. g. by threatening recourse to the United Nations or to the International Court of Justice, an economic or communication boycott, adoption of unfriendly attitudes in international negotiations, closing down of military bases, nationalisation, reduction of economic aid,

---

1) Lake Constance; boundary waters of Germany and the Netherlands.

etc.). Such an instrument is not solely applicable to transfrontier pollution and instead may be used on all kinds of occasion by the two countries, ultimately leading to results which are far from economically efficient, especially when the parties are unequal in status or exert very uneven degrees of pressure. While it hardly seems an appropriate instrument for permanently solving transfrontier problems in a cooperative spirit, it is a method which has already been used in conjunction with the pressure of public opinion in extreme pollution cases or in order to set negotiations in train.

Denouncing an agreement

The threat by a country to denounce an agreement on pollution because the other country has failed to comply with it is a means of pressure which may prove effective in the event that common resources are polluted, but not in the case of one-way transfrontier pollution.

Conclusions with regard to threshold instruments

Use of such an instrument as damages for transfrontier pollution is in line with traditional State practice and with generally accepted international law concepts. The advantages evident in case pollution levels are vastly exceeded, but if these are only slightly overshot the instrument would hardly seem appropriate. In the latter event it might be better advised to adopt the pollution charge, which does not have the contentious character associated with the more legal concept of damage payments and enables the substantive question of State liability to be avoided.

When pollution is very extensive it would no doubt be well-advised to provide for an insurance scheme and grant the same rights to both foreign and national victims.

Generally speaking, it might be well to set up a system for effectively and promptly settling disputes in order to promote the achievement of a negotiated solution, under the threat to resort to some organ which might impose a solution.

The instrument combining compensation for additional treatment with compensation for excessive pollution would seem to offer a better solution than damages, an added advantage being that the real cost must be ascertained by the polluted country when damage is slight.

Instruments or means of a non-economic nature (the pressure of public opinion and diplomatic pressure) do not guarantee the utmost efficiency and do not always have a sufficient preventive action when they are used only after an undesirable instance has taken place. They may be used as a last recourse when countries refuse to choose some other transfrontier pollution instrument. If some agency is set up to collect, compare and publish objective information on transfrontier pollution, the pressure instruments can then be used more evenly.

CLASS III. DISCRETE INSTRUMENTS

The mark of discrete instruments is an associated zero cost below a certain level which suddenly becomes non-zero and is constant (or increases) in value beyond that level. In order for a discrete instrument to be effective the associated cost must be such as to induce the party subject to the instrument not to exceed the threshold level. Typical discrete instruments are fines imposed on the polluter in case of excessive pollution, decisions taken by court or administrative authorities to cease operation or relocate polluting or pollution-sensitive activities, and physical measures taken to deal with sources of pollution.

Fines (Figure 11)

Payment of a fine by a country to a body outside the two countries affected by a transfrontier pollution episode does not seem to be an acceptable solution, since countries generally do not admit the power of some outside agency to impose and collect fines.

Figure 11
FINE

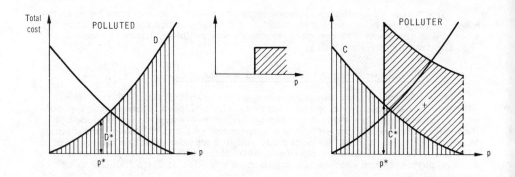

Figure 12
PURCHASE OF TREATMENT SERVICE/FINE

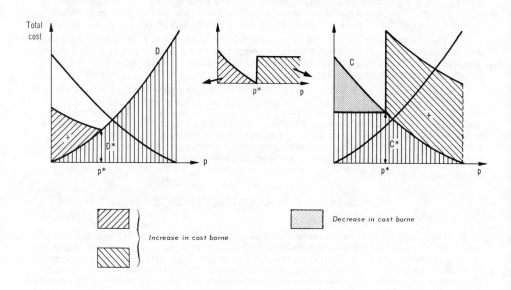

Fines so collected will be an efficient instrument only if their amounts are adequate compared to the gain which may result through lack of treatment or the chances of escaping the fine. A better monitoring system and the obligation to report any abnormal pollutant emission to the authorities are measures capable of strengthening such an instrument.

Countries might be more readily inclined to accept that any fines imposed to polluters of a country for non-compliance with certain regulation be transferred to the country which has suffered the damage. Such a provision might be implemented by transferring the fines to an international basin agency which would remit them to the polluted country.

In addition to being imposed by national bodies, fines might also be levied by international bodies common to the countries (such as the Court of Justice of the Communities in the case of non-compliance with Community rules).

Other deterrents

By compelling the party presumed to be at fault to deposit surety or bond (without interest) pending the settlement of fines and damages as well as paying court costs (including expenditure committed by the other party), the deterrent effect of fines can be strengthened and another discrete instrument thus be obtained. Such an instrument is only applicable pending a decision from a body whose jurisdiction is recognised by both countries.

Orders to cease operation or relocate certain activities

When some polluting enterprise is ordered to cease operation and hence must interrupt or sometimes relocate its activities, it is subjected to a cost unrelated to the level of pollution and having much the effect of a fine. This instrument may also be used to deal with pollution-sensitive activities (zoning imposing a residence ban). Such orders follow upon administrative or judicial decisions of a national character or of a body common to the countries. In transfrontier pollution a noteworthy precedent is the case of "Trail Smelter", which had to change its activities as a result of arbitration. It however seems unlikely that countries would yet be prepared to assign the right of closing or relocating an establishment to some joint body set up for dealing with transfrontier pollution questions (social consequences coming under the jurisdiction of the political authority).

In certain cases, however, the countries must first obtain permission from a joint body before initiating actions liable to cause transfrontier pollution (Convention of 1922 between Denmark and Germany, International Joint Commission for the Great Lakes, Commission of the European Communities for particularly dangerous nuclear experiments).

Direct action

Direct action consists in taking (or threatening to take) physical measures against sources of pollution located outside national territory. Such measures may consist in forcing a sea-going vessel to withdraw, and include stop and search, capture or destruction of the polluting vessel, thus constituting a threat to the ship-owner inducing a change in his behaviour.

Such an instrument may be implemented by extending, where pollution is concerned, the area at sea over which a country exercises jurisdiction either unilaterally (national legislation (1)) or by and international agreement (2) and by granting countries threatened with pollution in case of accident the right to take any protective measures (including that of blowing up the wreck (3)).

---

1) The United States can take action against ships on the high seas which constitute a danger for the shoreline (Federal Water Pollution Control Act, 1969). Canada has decided that ships entering the Arctic Ocean less than 100 nautical miles from its coasts must meet certain criteria to prevent pollution.

2) IMCO Convention (1954-62-69).

3) Brussels Convention, 1969.

Direct action would seem difficult in the case of transfrontier pollution from land sources but is an instrument which has already been accepted in dealing with the transfrontier pollution of coasts by ships. In this case some States have concluded treaties of mutual assistance to mitigate the effects of accidents (North Sea Agreement, 1969).

Conclusions regarding discrete instruments

Discrete instruments do not provide a graduated response to transfrontier pollution and are therefore bound to be less easily accepted by countries. In certain cases they are however used when the area of some State's jurisdiction is broadened, as an accessory to another instrument or as a protective measure. As a rule they would hardly seem suitable for solving transfrontier pollution problems, but in meeting certain special cases they should not be neglected.

TRANSFRONTIER POLLUTION COST-SHARING

Note by the Secretariat

INTRODUCTION

Transfrontier pollution is one of the international aspects of the environment. By transfrontier pollution is meant any discharge of matter (or release of energy), in the territory of one country, which is propagated by natural forces and causes damage in the territory of another country. It includes the pollution of international rivers and lakes, air space, and coastal waters in frontier areas.

In this paper we shall examine some economic principles which would make it possible to ensure that the environment was preserved and improved by joint action taken by the countries concerned, inspired by considerations of economic efficiency. To be more precise, we shall try to find rules for defining the costs countries must bear when there is transfrontier pollution which does not cause serious damage. We shall assume that the countries concerned have neither accepted the principle of full internalisation nor adopted its reverse and that consequently the polluted country and the polluting country must accept to bear some of the costs in connection with transfrontier pollution. We shall assume that the countries concerned have solved the problems of equity by negotiation.

SUMMARY AND CONCLUSIONS

It is well known that, in the case of pollution inside a country, the difference between the social cost and the private cost of polluting activities gives rise to a misallocation of resources and that the inability of the market mechanism to take account of these externalities can be overcome, as Pigou has shown, by imposing a tax at a rate equal to the difference between the marginal social cost and the marginal private cost. When the cost of damage due to pollution is not known, it is necessary to lay down a standard level for pollution and make the polluters bear the cost of the measures required to have this standard complied with (the OECD's Polluter-Pays Principle). These two methods can be applied inside a given country because there are numerous instruments for redistributing wealth with which to correct such of their effects as are politically or socially unfair.

In the case of transfrontier pollution one cannot safely assume that there are instruments which redistribute wealth between countries or between the polluters in one country and the victims of pollution in another, and one must be careful to distinguish between the problems of economic efficiency and those of ensuring equitable treatment.

If the countries concerned have solved the problem of equitable treatment, they will be able to choose principles for dealing with transfrontier pollution which are conducive to economic efficiency. When the waste treatment costs and the damage costs are well known and the countries concerned agree in their estimates of them, the quest for economic efficiency in dealing with transfrontier pollution raises no new problem, but in practice, as countries seldom agree in their estimates of all these costs and there is no authority who can give a ruling, the situation has to be further investigated with due regard for the particular nature of the relations between the countries concerned.

The conclusion reached by the theoretical approach in Annex 1 is that the most satisfactory options are:

a) the modified civil liability principle (M. C. L. P.) and

b) the modified victim-pays principle (M. V. P. P.).

The modified civil liability principle (M. C. L. P.) is that the polluting country shall be financially responsible for the emissions of pollutants from its territory as abated by the damage cost in the polluted country corresponding to the agreed level of pollution, and that the polluted country shall bear the cost of the damage done in its territory by pollution at that agreed level. The modified victim-pays principle (M. V. P. P.) is that the polluted country shall be financially responsible for the emissions of pollutants from the territory of the polluting country as reduced by the waste treatment cost to the agreed standard, and that the polluting country shall bear the waste treatment cost to that standard or, if the standard is exceeded, shall be financially responsible for its emissions of pollutants as reduced by the cost of the damage done in the polluted country by pollution at the agreed level.

These two principles are in line with the polluter-pays principle, since the polluter pays the treatment costs down to the agreed level of pollution. They differ according to whether the optimum level of pollution is chosen by the polluting country (M. C. L. P.) or by the polluted country (M. V. P. P.) and according to whether the gain in efficiency as compared with the agreed level accrues to the polluting country (M. C. L. P.) or to the polluted country (M. V. P. P.).

When the cost of pollution control is not properly known and the damage cost is well known, the modified civil liability principle obliges the polluting country to make a political or economic choice which determines the cost of the pollution control measures it will take in its territory. On the other hand, when the damage cost is not properly known, but the cost of pollution control is well known, the modified victim-pays principle obliges the polluted country to make a political or economic choice which determines the damage cost which it alone has to bear. One can therefore choose between the two principles, depending on which costs turn out to be the less known.

If the damage cost is unknown to both countries and they set themselves pollution control targets, they can lay down optimum quantities for emissions into the environment and calculate the damage costs approximately, on which case either of the two principles may be used.

The reason why there is a choice between the two principles in the case of transfrontier pollution is that one or the other country is able to minimise one of the elements in its total costs and seek the economic optimum.

In the case of pollution inside a country, the victims of pollution are scattered and seldom have an organisation to represent them effectively, so that it has seemed unlikely that they could assume direct responsibility for seeking the optimum solution. In domestic policy, therefore, Governments have usually shown a preference for the modified civil liability principle, but the same need not necessarily apply at international level when there are only a few countries involved.

As a rule damage costs are apparently not so well known as pollution control costs, at least in the short term, so that it would seem preferable to choose the modified victim-pays principle.

Annex 1

ECONOMIC ANALYSIS OF COST-SHARING

SOME FINDINGS OF THE SEMINAR ON TRANSFRONTIER POLLUTION

The question of sharing the costs arising from transfrontier pollution between the polluting country and the polluted country has been examined in a number
of theoretical economic studies submitted to the Seminar on Transfrontier Pollution (1). In particular, Muraro and Smets have examined in detail some of the
advantages and disadvantages of systems in which the polluting country bears some
or all of the waste treatment cost, and of systems in which it bears some or all of
the waste treatment cost and the damage cost. The analysis given below is the
logical follow-up to these studies. Like the report by Scott, it lays stress on defining countries' responsibilities for the optimum management of a common resource in cases where pollution often causes damage to a public good.

ECONOMIC ANALYSIS MODEL

We shall restrict ourselves to the very simple case of transfrontier pollution
in which two countries pollute a shared environment and consequently suffer damage. We shall assume that the result of the transfrontier pollution is to reduce
the welfare of the two countries by a certain overall amount T^* and that one of the
objectives of their Governments is to maximise the welfare of the two countries.
If the pollution control measures affect only the total amount T, the objective will
be attained by implementing the combination of measures which minimises it. If
the simultaneous development of activities which are polluting and/or sensitive to
pollution should be a satisfactory course of action, a necessary condition will be
for each country to have greater welfare than it had in the days before transfrontier
pollution. One will then have to find out how the welfare of each country has varied
before the optimum pollution control measures were introduced and after. If both
countries are improving their welfare, they will readily accept the optimum solution. Such will often be the case when a lake is being polluted by two countries.
However, the welfare of the polluting country may well be reduced if it has to take
stricter measures against pollution. This will often happen when the two countries
lie along a river and one of them pollutes the other. In this situation the countries
will encounter difficulties in implementing the optimum solution and negotiations
will be necessary, for instance, to define the maximum amount of the upstream
country's loss of welfare.

More generally speaking, the minimum overall cost T^* (1) corresponding to
the optimum strategy should be shared between the two countries. Each country
will bear the cost of damage caused by pollution when the optimum anti-pollution
measures are taken (residual damage) and also a part of the cost of the pollution
control measures. As a rule it will rarely be necessary for a country to pay an
indemnity to the other country also, but it will often be necessary for one country
to finance some of the pollution control expenditure incurred in the other country.
Therefore the minimum overall cost T^* will be broken down into a cost T_1^* borne
by country 1 and a cost T_2^* borne by country 2. Calculating the formula according

1) See Problems in Transfrontier Pollution, OECD, Paris, 1974.
2) See list of symbols at the end of this Annex.

to which the minimum overall cost T^* will be shared is a problem of equity which is dealt with in other reports in this book.

COST-SHARING

From the economic analysis it will be possible to suggest a cost-sharing formula for the overall cost T, once the cost-sharing formula for the minimum overall cost T^* is known, insofar as the countries can determine the costs of pollution control and the damage costs (1).

If countries 1 and 2 decide to bear the costs T_1 and T_2, and if T_1 and T_2 are such that

a) $T_1 + T_2 = T$

b) T_1 and T_2 are positive

c) either T_1 and T_2 have a single minimum at the optimum, or one of the two functions is constant and the other has a single minimum at the optimum,

d) the value of T_1 at the optimum is T_1^*

e) the value of T_2 at the optimum is T_2^*,

it will then be advantageous to both countries to select the optimum.

For example, the countries might share between them the minimum overall cost T^* and the difference in cost $(T-T^*)$ from the optimum situation (2).

The overall cost T will be the sum of the cost of the pollution control measures C and the damage cost due to pollution in each country $D = T = C + D$

---

1) We shall examine in Annex 3 the case where the countries cannot find out the damage costs and have set pollution targets to be achieved simultaneously.

2) If countries 1 and 2 adopt the principle of shared responsibility, they will bear the costs

$$\begin{cases} T_1 = \alpha (T-T^*) + T_1^* = \alpha (T-T^*) + \beta T^* \\ T_2 = \alpha (1 - \alpha) (T-T^*) + T_2^* = (1 - \alpha) (T-T^*) + (1 - \beta) T^* \end{cases}$$

with $0 \leqslant \alpha \leqslant 1$ and $0 \leqslant \beta \leqslant 1$. This principle of shared responsibility is governed by the two parameters α and β which are not necessarily equal. The countries might for example, agree to share the overall minimum cost T^* in accordance with their national products, while allocating liability for the difference in cost from the optimum situation to the country responsible for this difference. Selecting an intermediate value for α will give the two countries similar responsibilities for finding the economic optimum. When α is equal to 0 or 1, one of the countries will bear the entire responsibility for minimising the overall cost T and the other country will not be affected by the first country's choice.

If one allocates entire responsibility to a single country, it should be possible to reduce divergencies from the optimum, so that this course appears preferable. Thus when $\alpha = \frac{1}{2}$ and the countries take action only when the cost difference exceeds A, they will each accept the cost difference $\frac{1}{2} (T-T^*) = A$, and the overall welfare will be reduced by 2 A. On the other hand, if $\alpha = 1$ or $\alpha = 0$, the two countries will only accept the cost difference $T-T^* = A$. Total welfare will be reduced by only A and the welfare of the country which is not responsible for minimising the total cost T will remain unchanged.

If the cost of pollution control measures C corresponds to measures costing C_1 in country 1 and measures costing C_2 in country 2, and if the damage cost D consists of the cost D_1 in country 1 and the cost D_2 in country 2, the overall cost will be

$$T = C_1 + C_2 + D_1 + D_2$$

If the asterisk denotes the value at the economic optimum, the overall cost can also be written:

$$T = (C_1 - C_1^*) + (C_2 - C_2^*) + (D_1 - D_1^*) + (D_2 - D_2^*) + T_1^* + T_2^*$$

and countries 1 and 2 will bear, for example,

$$\begin{cases} T_1 = \alpha \, (C_1 - C_1^*) + (1 - \beta)\,(C_2 - C_2^*) + \gamma\,(D_1 - D_1^*) + (1 - \delta)\,(D_2 - D_2^*) + T_1^* \\ T_2 = (1 - \alpha)\,(C_1 - C_1^*) + \beta\,(C_2 - C_2^*) + (1 - \gamma)\,(D_1 - D_1^*) + \delta\,(D_2 - D_2^*) + T_2^* \end{cases}$$

where α, β, γ, and δ are four parameters having values between zero and unity.

We are concerned here only with choosing parameters such α, β, γ, and δ and not with choosing the costs T_1^* and T_2^* borne by the two countries when the optimum pollution control strategy is brought to bear.

If the pollution control measures which country 1 is able to implement reach an intensity or degree $(X_i - x_i)$, the cost of these measures will be:

$$C_1\,(X_i,\ x_i)$$

where X_i is the initial value and x_i the chosen value of the variable in question.

Similarly, country 2 can implement measures which reduce the variables Y_j to the value y_j and cost $C_2(Y_j, y_j)$. The effect of these measures will be to reduce the damage cost from $D_1\,(X_i, Y_j)$ to $D_1(x_i, y_j)$ in country 1 and from $D_2(X_i, Y_j)$ to $D_2(x_i, y_j)$ in country 2.

In general the variables x and y correspond to pollution control measures, while the variables X and Y relate to the socio-economic situation or to investment and are therefore less subject to variation. The overall economic optimum is reached by maximising the welfare with regard to the 4 variables X, Y, x and y. Here we are concerned only with the choice of the variables x and y which describe the environmental policy and not with the variables X and Y which describe other policies of the Governments (policies for regional development, urbanisation, industrialisation, etc.).

For instance, let X and Y be the values for gross pollution (amounts of pollutant produced) in countries 1 and 2, and x and y the figures for net pollution (amounts of pollutant discharged into the environment, or residual pollution after treatment) in countries 1 and 2. The waste treatment costs will increase with the amounts of pollutant eliminated X - x, Y - y and the damage will be proportional to the pollution levels p_1 and p_2 in countries 1 and 2:

$$p_1 = a_{11}x = a_{12}y$$

$$p_2 = a_{21}x + a_{22}y$$

This case is examined in detail in Annexes 2 and 3.

The quantities x and y might also stand for the populations exposed to pollution p. The costs of displacing X - x and Y - y persons will be $C_1(X-x)$ and $C_2(Y-y)$ respectively and the damage will be proportional to the numbers of persons, x and y, exposed to the pollutant.

An asterisk is used to indicate the values taken by the variables x_i and y_j when the overall cost $T(x_i, y_j)$ is a minimum and is equal to $T^*(x_i^*, y_j^*)$. (1)

---

1) x_i^* and y_j^* are solutions of

$$\begin{cases} \dfrac{\partial C_1}{\partial x_i} + \dfrac{\partial D_1}{\partial x_i} + \dfrac{\partial D_2}{\partial x_i} = 0 \\[2mm] \dfrac{\partial C_2}{\partial y_j} + \dfrac{\partial D_2}{\partial y_j} + \dfrac{\partial D_1}{\partial y_j} = 0 \end{cases}$$

if we leave out the special case where the optimum is not analytical (corner solution).

If country 1 alone controls the application of the measures expressed by the variable x_i, it will choose the optimum level x_i^* if it is to bear the additional cost of the pollution control measures $C_1(X_1, x_i) - C_1(X_i, x_i^*)$, the additional damage cost within its own territory $D_1(x_i) - D_1(x_i^*)$ and the additional damage cost in the neighbouring country $D_2(x_i) - D_2(x_i^*)$. If country 1 is not responsible for damage in the neighbouring country, it will generally choose the wrong level for x_i.

If country 1 is responsible for a fraction α of the additional cost $C_1(X_i, x_i) - C_1(X_i, x_i^*)$, it must also be responsible for the same fraction of the damage cost within its own territory and in the neighbouring country. Since it appears preferable for all responsibility for action expressed by the variable x_i to be placed on a single country so as to make sure that its impact is not negligible and that the country concerned takes quick action in the absence of optimisation, a choice will have to be made between the system in which country 1 is responsible for damage and pollution control measures ($\alpha = 1$) and the system in which country 2 accepts this same responsibility ($\alpha = 0$). The problem of cost-sharing in the case of one-way pollution is illustrated in figures 1, 2 and 3.

COMPARISON BETWEEN COST-SHARING SYSTEMS

The first system will always be valid, since the pollution control measures in question will be taken in country 1. The second system means that country 1 will agree to take measures to an extent to be decided by country 2 and whose cost will be borne by country 2. This solution will not be acceptable to country 2 if country 1 can alter this cost /i. e. if $C_1(X_i, x_i) - C_1(X_i, x_i^*)$ depends on X_i, this being a variable which depends only on country 1, such as the level of industrial activity/. If x_i is the net pollution (quantity of pollutant discharged) and X_i is the gross pollution (quantity of a pollutant produced), country 1 will be able to alter the cost borne by country 2 by raising the level of output of the enterprises originating the pollutant. Country 2 will not accept this system if it means that the effects of the net pollution by country 1 on country 2 may vary unpredictably, and country 2 will then demand that country 1 be held responsible for any unexpected discharge of pollutants (filtration system out of action, operational error, etc.). In these conditions we shall be comparing the first system with the second system as corrected by the first system if agreed levels are exceeded. We shall assume that both countries are equally capable of determining an economic optimum.

If the estimated additional pollution control costs $C_1(x_i) - C_1(x_i^*)$ and additional damage costs $D_1(x_i) - D_1(x_i^*)$ are satisfactory to both countries, but neither can agree on the damage cost $D_2(x_i)$ imposed by one country on the other country (e. g. if the polluted country puts a higher value on it because of social preferences which are not recognised by the polluting country), it will be preferable to choose the second system in which country 2 has to reveal, by making an economic or political choice, what value it actually does put on the reciprocal damage cost $D_2(x_i)$. If the damage caused by one country in another country shows a marked increase, it may be assumed that the exceptional damage cost can be fairly reliably estimated, because the quantifiable kinds of degradation will overshadow the kinds which are more difficult to quantify at tolerable pollution levels (e. g. loss of amenities). There will thus be no difficulty in calculating compensation for the additional damage when pollution exceeds the agreed level. On the contrary, if both countries agree in their estimates of the reciprocal damage $D_2(x_i)$, but cannot agree on the additional pollution control costs $C_1(x_i) - C_1(x_i^*)$ and/or the additional damage costs $D_1(x_i) - D_1(x_i^*)$, it will be preferable to choose the first system in which country 1 has to reveal, by making an economic or political choice, the true value of the additional cost of the measures expressed by the variable x_i: $C_1(x_i) - C_1(x_i^*)$ and of the additional damage costs $D_1(x_i) - D_1(x_i^*)$.

If the damage cost functions are separable (1) (and in particular if the damage is a linear function of pollution):

---

1) The following are the cost-sharing formulas under the first system when these damage cost functions are non-separable:

a) $\begin{cases} T_1 = C_1(x) - C_1(x^*) + D_1(x, y) - D_1(x^*, y) + D_2(x, y) - D_2(x^*, y) + T_1^* \\ T_2 = C_2(y) - C_2(y^*) + D_1(x^*, y) - D_1(x^*, y^*) + D_2(x^*, y) - D_2(x^*, y^*) + T_2^* \end{cases}$

$$\begin{cases} D_1(x,y) = D_{11}(x) + D_{12}(y) \\ D_2(x,y) = D_{21}(x) + D_{22}(y) \end{cases}$$

The cost-sharing formula for the first system is:

$$\begin{cases} T_1 = C_1(x)-C_1(x^*)+D_{11}(x)-D_{11}(x^*)+D_{21}(x)-D_{21}(x^*)+T_1^* \\ T_2 = C_2(y)-C_2(y^*)+D_{12}(y)-D_{12}(y^*)+D_{22}(y)-D_{22}(y^*)+T_2^* \end{cases}$$

This formula can also be written:

$$\begin{cases} T_1 = C_1(x)+D_{11}(x)+D_{21}(x)-D_{21}(x_c)+D_{12}(y_c) \\ T_2 = C_2(y)+D_{22}(y)+D_{12}(y)-D_{12}(y_c)+D_{21}(x_c) \end{cases}$$

This means that country 1 undertakes to bear the cost of the measures required to bring the variable x (e. g. the pollution level) down to the value x_c and the cost of the damage done in country 1 by country 1 when $x = x_c$: $C_1(x_c)+D_{11}(x_c)$, as well as the damage cost in country 1 when country 2 has brought the variable y down to the level y_c: $D_{12}(y_c)$. Meanwhile country 2 undertakes at the same time to bear the cost $C_2(y_c)+D_{22}(y_c)$ and the residual damage cost $D_{21}(x_c)$.

This is a case of applying the modified civil liability principle by not paying compensation for the residual damage, since country 1 is responsible for the additional damage cost which it imposes on country 2, viz. $D_{21}(x)-D_{21}(x_c)$, and country 2 for the additional damage cost which it imposes on country 1, viz. $D_{12}(y)-D_{12}(y_c)$. The polluting countries pay for reducing the levels x and y from the agreed levels x_c and y_c to the optimum levels x^* and y^*.

When the marginal damage cost imposed by country 1 on country 2 increases as a result of action by country 2, the cost T_1 borne by country 1 increases, whereas when the cost C_1 of pollution control measures increases in country 1 as a result of action by country 1, the cost T_2 borne by country 2 remains the same. So it would appear that an agreement based on the modified civil liability principle is not altered unfairly in cases in which the costs C_1 and C_2 of pollution control measures are altered, but that it is less satisfactory when the reciprocal damage costs D_{21} and D_{12} are altered (the agreement remains equitable over time when the variables outside the model change).

If the damage cost is separable, the cost-sharing formula for the second system will be :

Note 1) of the page 93 (cont'd) :

b) $$\begin{cases} T_1 = C_1(x)-C_1(x^*)+D_1(x,y^*)-D_1(x^*,y^*)+D_2(x,y)-D_2(x^*,y)+T_1^* \\ T_2 = C_2(y)-C_2(y^*)+D_1(x,y)-D_1(x,y^*)+D_2(x^*,y)-D_2(x^*,y^*)+T_2^* \end{cases}$$

c) $$\begin{cases} T_1 = C_1(x)-C_1(x^*)+D_1(x,y)-D_1(x^*,y)+D_2(x,y^*)-D_2(x^*,y^*)+T_1^* \\ T_2 = C_2(y)-C_2(y^*)+D_1(x^*,y)-D_1(x^*,y^*)+D_2(x,y)-D_2(x,y^*)+T_2^* \end{cases}$$

d) $$\begin{cases} T_1 = C_1(x)-C_1(x^*)+D_1(x,y^*)-D_1(x^*,y^*)+D_2(x,y^*)-D_2(x^*,y^*)+T_1^* \\ T_2 = C_2(y)-C_2(y^*)+D_1(x,y)-D_1(x,y^*)+D_2(x,y)-D_2(x,y^*)+T_2^* \end{cases}$$

These four different formulas reflect the fact that the effects (damage) cannot be separated from their causes (e. g. the pollution levels). In some cases the effect is synergic or results from congestion or over-crowding. We shall not deal here with the problem of choosing the best formula for finding the optimum solution by allowing the two countries to make their own successive decisions, as this would involve us in complex mathematical calculations from which it would be difficult to deduce simple principles. The cost-sharing formulas are the same for the second system as for the first system, except that country 1 bears T_2 and country 2 bears T_1.

$$\begin{cases} T_1 = C_2(y) - C_2(y^*) + D_{12}(y) - D_{12}(y^*) + D_{22}(y) - D_{22}(y^*) + T_1^* \\ T_2 = C_1(x) - C_1(x^*) + D_{11}(x) - D_{11}(x^*) + D_{21}(x) - D_{21}(x^*) + T_2^* \end{cases}$$

This formula can also be written:

$$\begin{cases} T_1 = C_2(y) - C_2(y_c) + D_{12}(y) + D_{22}(y) - D_{22}(y_c) + C_1(x_c) + D_{11}(x_c) \\ T_2 = C_1(x) - C_1(x_c) + D_{11}(x) - D_{11}(x_c) + D_{21}(x) + C_2(y_c) + D_{22}(y_c) \end{cases}$$

This means that country 1 undertakes to bear the cost of the measures required to bring the variable x down to the value x_c: $C_1(x_c) + D_{11}(x_c)$ and the residual damage cost $D_{12}(y_c)$ in the country 1 when country 2 has brought the variable y down to level y_c, while country 2 undertakes at the same time to bear the cost $C_1(y_c) + D_{22}(y_c)$ and the residual damage cost $D_{21}(x_c)$.

When level x_c (or y_c) is chosen so as to minimise the cost of the pollution control measures taken inside country 1 (or 2) and the damage cost in country 1 (or 2), viz. $C_1(x) + D_{11}(x) + D_{12}(y)$ /or $C_2(y) + D_{22}(y) + D_{21}(x)$_7, i.e. when one leaves out the reciprocal damage costs D_{12} and D_{21}, both countries will reveal the values of their marginal damage costs

$$\left[\frac{\partial D_{11}}{\partial x} \right]_{x_c} \quad \text{and} \quad \left[\frac{\partial D_{22}}{\partial y} \right]_{y_c}$$

and from these one can obtain an estimate of the additional damage costs $D_{11}(x) - D_{11}(x_c)$ and $D_{22}(y) - D_{22}(y_c)$ financed by countries 2 and 1 in respect of damage which occurs physically in countries 1 and 2.

Under the second system, country 1 will ask country 2 to treat its wastes to level y^* so as to reduce the effects in country 1 of the pollution discharged in country 2, and country 2 will ask country 1 to treat its wastes to level x^*. This is really the modified victim-pays principle, since the polluted countries agree to pay for reducing the variables x and y from levels x_c and y_c to levels x^* and y^*. The victim-pays principle can only be used if both countries take the measures necessary to reach the agreed levels x_c and y_c.

If both countries do not meet their commitments regarding the agreed levels, i.e. if country 1 does not reduce x to level x_c and/or country 2 does not reduce y to level y_c, the country which fails to meet its commitments must bear the consequences (breach of an agreement) and indemnify the other country so that the latter does not have to meet higher costs than those which it had undertaken to bear.

When a country alters the reciprocal damage cost (D_{12} or D_{21}), the expenditure to be met by that country will alter correspondingly. For example, if a country uses a process which is more sensitive to pollution, it will have to bear alone the additional costs arising from lowering the optimum level of pollution, whereas if a country increases the cost of pollution control, as for example by developing a polluting industry, it will have to bear alone the additional cost of keeping its discharges down to the agreed level. The other country will then have no increased costs to bear if the additional cost of treatment is independent of the level of output, and will only have to meet a slight increase if it is not. Thus an agreement based on the modified victim-pays principle can remain equitable both in cases where the reciprocal damage cost is altered and in cases where the cost of pollution control is altered (the agreement remains equitable over time when the outside variables change).

In conclusion, either the modified civil liability principle or the modified victim-pays principle may be made the basis for sharing out the global costs due to transfrontier pollution. In cases where the costs of pollution control are not as well known as the damage costs, the modified civil liability principle might be chosen, while in cases where the costs of pollution control are the better known, preference will be given to the modified victim-pays principle.

Figure 1

FIRST EXAMPLE OF COST SHARING :
THE PRINCIPLE OF MODIFIED CIVIL LIABILITY

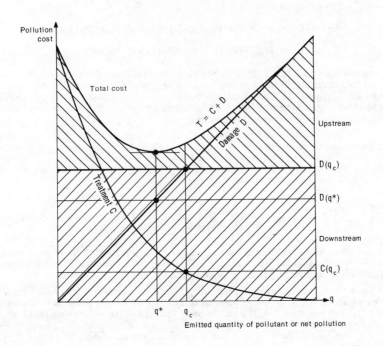

Emitted quantity of pollutant or net pollution

Model

Source of pollution in the upstream country only.

Cost of waste treatment to reduce pollutant discharges to q tonnes: C(q) in the upstream country, 0 in the downstream country.

Cost of damage due to q tonnes of pollutant discharged into the environment: 0 in the upstream country and D(q) in the downstream country.

Cost borne by the downstream country: $D(q_c)$, i.e. the damage cost resulting from the discharge of q_c tonnes by the upstream country or, if the upstream country discharges quantity q, the damage cost D(q) less the transfer $D(q) - D(q_c)$ /compensation for the damage $(q > q_c)$ or contribution to the additional waste treatment costs $(q < q_c)$/.

Therefore the level of discharges into the environment will make no difference to the downstream country.

Cost borne by the upstream country: the difference between the total cost T(q) = C+D and the cost borne by the downstream country $D(q_c)$.

Discharge level chosen by the upstream country: q^* in order to minimise the cost to be borne.

Note: The downstream country will transfer $D(q_c) - D(q^*)$ to the upstream country.

Figure 2

SECOND EXAMPLE OF COST SHARING :
THE MODIFIED VICTIM-PAYS PRINCIPLE

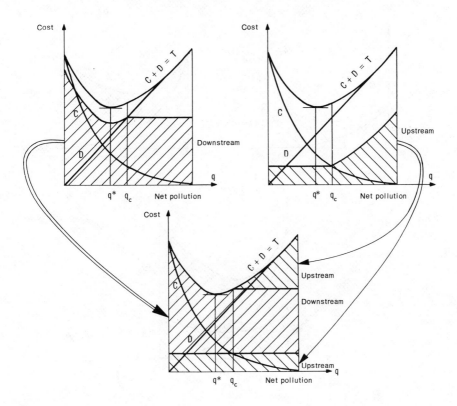

Model: see Figure 1

Cost borne by the upstream country: The cost of waste treatment $C(q_c)$ to reach the discharge level q_c, supplemented if $q > q_c$ by the excess damage cost $D(q) - D(q_c)$. The upstream country will try not to exceed the level q_c and will be unaffected by the choice of discharge level if $q < q_c$.

Cost borne by the downstream country: The difference between the total cost $T(q) = C + D$ and the cost borne by the upstream country, i.e. $T - C(q_c)$ if $q < q_c$. The downstream country will receive compensation $D(q) - D(q_c)$ if $q > q_c$.

Discharge level chosen by the downstream country: q^* in order to minimise the cost to be borne.

Note: The downstream country will transfer $D(q_c) - D(q^*)$ to the upstream country.

Figure 3

THIRD EXAMPLE OF COST-SHARING :
THE PRINCIPLE OF SHARED RESPONSIBILITY

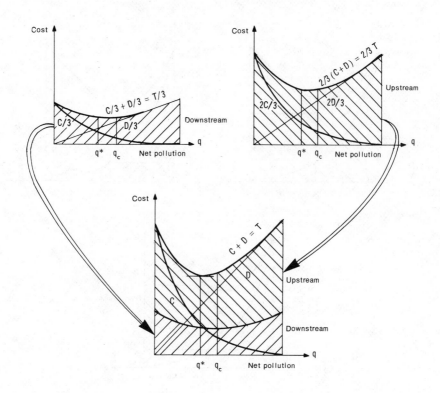

Model: see Figure 1.

Cost borne by the downstream country: $\frac{1}{3}$ (C + D)

Cost borne by the upstream country: $\frac{2}{3}$ (C + D)

Discharge level chosen by the upstream and downstream countries: q^* in order to minimise the costs to be borne.

Note: The upstream country transfers $\frac{1}{3}$ of the damage cost: $\frac{1}{3}$ D(q^*) to the downstream country, which transfers $\frac{2}{3}$ of the waste treatment cost $\frac{2}{3}$ C(q^*) to the upstream country. The net transfer to the upstream country will be $\frac{2}{3}$ C(q^*) - $\frac{1}{3}$ D(q^*).

T	Loss of welfare or "overall cost" to all the countries concerned due to transfrontier pollution ($T = C + D$)
T^{*}	Minimum value of T when the cheapest pollution control strategy is applied in the optimum way (optimum strategy)
T_1, T_2	Loss of welfare in countries 1 and 2, or cost borne by countries 1 and 2 as a result of transfrontier pollution ($T = T_1 + T_2$)
T_1^{*}, T_2^{*}	Minimum value of T_1 and T_2 when the optimum strategy is adopted ($T^{*} = T_1^{*} + T_2^{*}$)
C	Global cost of pollution control measures
C_1, C_2	Cost of pollution control measures taken in countries 1 and 2 (but not necessarily borne by countries 1 and 2) ($C = C_1 + C_2$)
C_1^{*}, C_2^{*}	Cost of pollution control measures taken in countries 1 and 2 in line with the optimum strategy ($C^{*} = C_1^{*} + C_2^{*}$)
D	Total damage cost due to transfrontier pollution
D_1, D_2	Damage cost in countries 1 and 2 ($D = D_1 + D_2$)
D^{*}	Total damage cost when the optimum strategy is adopted
D_1^{*}, D_2^{*}	Damage cost in countries 1 and 2 when the optimum strategy is adopted ($D^{*} = D_1^{*} + D_2^{*}$)
x_i, y_i	Value of the variables related to pollution control measures whose cost is nil when $x_i = X_i$, $y_i = Y_i$, and is positive when $x_i < X_i$ and $y_i < Y_i$ ($\frac{\partial C}{\partial x_i} \leqslant 0$, $\frac{\partial C}{\partial y_i} \leqslant 0$, $\frac{\partial D}{\partial x_i} \geqslant 0$, $\frac{\partial D}{\partial y_i} \geqslant 0$) ($x_i$ in country 1 and y_i in country 2)
x_i^{*}, y_i^{*}	Values of x_i and y_i when the optimum strategy is adopted ($\frac{\partial T}{\partial x_i} = \frac{\partial T}{\partial y_i} = 0$ when $x_i = x_i^{*}$, $y_i = y_i^{*}$)
x_{ic}, y_{ic}	Agreed values for x_i and y_i ($0 \leqslant x_{ic} \leqslant X_i$, $0 \leqslant y_{ic} \leqslant Y_i$)

N.B. Except in very special cases,

$T_1 \neq C_1 + D_1$ and $T_2 \neq C_2 + D_2$

but $T = C + D = C_1 + D_1 + C_2 + D_2 = T_1 + T_2$

Annex 2

ANALYSIS OF A PARTICULAR CASE OF TRANSFRONTIER POLLUTION

The purpose of this Annex is to describe and discuss the merits of the various cost-sharing methods in the particular case in which the best way of controlling pollution is to keep emissions into the environment under control (by filtration, purification and other treatment). Our analysis will be based on mathematical concepts and the usual optimisation techniques.

DESCRIPTION OF MATHEMATICAL MODEL

Let us take the problem raised by discharges of pollutants from two countries 1 and 2 into an environment which they share and let us assume that the damage cost suffered by each country is an increasing function of the level of pollution in it (p_1 or p_2):

$$\text{Country 1: } D_1 = D_1(p_1), \quad \frac{dD_1}{dp_1} \geqslant 0 \tag{1}$$

$$\text{Country 2: } D_2 = D_2(p_2), \quad \frac{dD_2}{dp_2} \geqslant 0 \tag{2}$$

The damage cost will be nil below the "nil effect" level and infinite above the "lethal" level. The pollution level p_i in each country will depend on the quantities of pollutant discharged by country 1 and country 2 into the environment, viz. q_1 and q_2, so that

$$p_1 = p_1(q_1, q_2) \tag{3}$$

$$p_2 = p_2(q_1, q_2) \tag{4}$$

and will be an increasing function of q_1 and q_2.

Let us make the simplifying assumption that each country can only control the quantities q_1 and q_2 by treating the effluents (individually or jointly) and cannot treat the environment as a whole, exert an effect on the damage cost function, or reduce the numbers of persons exposed to pollution or the level of activity of the polluting enterprises. Let us also assume that the implementation of an environmental policy will not affect the level of activity of the polluting enterprises. The minimum cost of bringing the discharges by country 1 down to level q_1 by suitable waste treatment in country 1 will then be $C_1(q_1)$ and the cost of waste treatment in country 2 will likewise be $C_2(q_2)$.

When both countries act independently of one another they will have to bear the total costs:

$$C_1(q_1) + D_1(p_1) \text{ and } C_2(q_2) + D_2(p_2).$$

The damage costs can always be expressed separately as the sum of two functions:

$$\begin{cases} D_1(p_1) = D_{11}(q_1) + D_{12}(q_1, q_2) & (5) \\ D_2(p_2) = D_{21}(q_1, q_2) + D_{22}(q_2) & (6) \end{cases}$$

with

$$\begin{cases} D_{12}(q_1, 0) = 0 \\ D_{21}(0, q_2) = 0 \end{cases}$$

The costs D_{11} and D_{22} will correspond to the damage which each country inflicts on itself and the costs D_{12} and D_{21} to the damage which the two countries inflict on one another. One-way transfrontier pollution will arise when the discharges by one country do not affect the other country (D_{12} or D_{21} is not a null function) and reciprocal transfrontier pollution will arise when the discharges by each country affect the other country (D_{12} and D_{21} are not null functions). If each country is only affected by its own discharges ($D_{12} = D_{21} = 0$) there will be no transfrontier pollution. In the case of one-way pollution, the "upstream" country will mean the country which is not affected by the discharges from the other country (called the "downstream country" (see Figure 1).

Figure 1

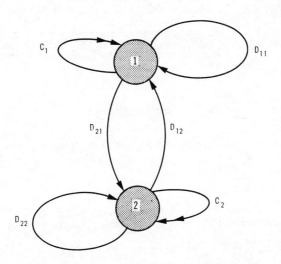

INDIVIDUAL OPTIMUM AND JOINT OPTIMUM

The levels of discharge which minimise the costs borne by each country will correspond to the individual economic optimum. Let us assume that these levels $\bar{q}_1$ and $\bar{q}_2$ are the only solutions to the equations

$$\begin{cases} \dfrac{\partial C_1}{\partial q_1} + \dfrac{\partial D_{11}}{\partial q_1} + \dfrac{\partial D_{12}}{\partial q_1} = 0 & (7) \\[2ex] \dfrac{\partial C_2}{\partial q_2} \quad \dfrac{\partial D_{22}}{\partial q_2} \quad \dfrac{\partial D_{21}}{\partial q_2} = 0 & (8) \end{cases}$$

Let us also assume that the conditions for the second derivatives are satisfied so as to enable the unique solution to be stable.

The levels of discharge which minimise the overall cost $T = C_1 + C_2 + D_1 + D_2$ will correspond to the joint economic optimum for both countries taken together. Let us assume that these levels q_1^* and q_2^* are the only solutions to the equations

$$\begin{cases} \dfrac{\partial T}{\partial q_1} = 0 \\[2mm] \dfrac{\partial T}{\partial q_2} = 0 \end{cases}$$

i.e.:
$$\begin{cases} \dfrac{\partial C_1}{\partial q_1} + \dfrac{\partial D_{11}}{\partial q_1} + \dfrac{\partial D_{12}}{\partial q_1} + \dfrac{\partial D_{21}}{\partial q_1} = 0 \qquad (9) \\[4mm] \dfrac{\partial C_2}{\partial q_2} + \dfrac{\partial D_{22}}{\partial q_2} + \dfrac{\partial D_{21}}{\partial q_2} + \dfrac{\partial D_{22}}{\partial q_2} = 0 \qquad (10) \end{cases}$$

Let us also assume that:

$$\frac{\partial^2 T}{\partial q_1^2} \; \frac{\partial^2 T}{\partial q_2^2} > \left[\frac{\partial^2 T}{\partial q_1 \, \partial q_2} \right]^2 \qquad (11)$$

so that the overall cost will be at a minimum at levels $q_1{}^*$ and $q_2{}^*$. Levels $\bar{q}_1$, $\bar{q}_2$ and $q_1{}^*$, $q_2{}^*$ will usually be different as a result of the damage done in one (or each) country by the discharges in the other country. The overall cost T at levels $q_1{}^*$ and $q_2{}^*$ will always be lower than the total cost T at levels $\bar{q}_1$ and $\bar{q}_2$. Generally speaking, it is found that pollution is less with the joint optimum and that the result of taking account of the damage caused to the other country is to increase the level of waste treatment and to reduce the levels of discharge.

SEPARABILITY

When the damage costs D_1 and D_2 are separable, i.e. when

$$D_{12}(q_1, \; q_2) = D_{12}(q_2) \qquad (12)$$

$$D_{21}(q_1, \; q_2) = D_{21}(q_1) \qquad (13)$$

the problem of reciprocal transfrontier pollution can be divided into two cases of simultaneous independent one-way transfrontier pollution whose costs are :

(a) $C_1(q_1)$, $D_{11}(q_1)$, $D_{21}(q_1)$

(b) $C_2(q_2)$, $D_{22}(q_2)$, $D_{12}(q_2)$

The first case concerns the emission by country 1 of a pollutant which affects countries 1 and 2, while the second case concerns the emission by country 2 of a pollutant which affects countries 2 and 1. Country 1 will solve the first of these problems by itself by choosing level $\bar{q}_1$ which is the solution of the equation

$$\frac{\partial C_1}{\partial q_1} = \frac{\partial D_{11}}{\partial q_1} = 0 \qquad (14)$$

and countries 1 and 2 will solve this first problem jointly by choosing level $q_1{}^*$ which is the solution of the equation

$$\frac{\partial C_1}{\partial q_1} + \frac{\partial D_{11}}{\partial q_1} + \frac{\partial D_{21}}{\partial q_1} = 0 \qquad (15)$$

Country 2 will solve the second problem by itself by choosing level $\bar{q}_2$ which is the solution of the equation

$$\frac{\partial C_2}{\partial q_2} + \frac{\partial D_{22}}{\partial q_2} = 0 \qquad (16)$$

and countries 2 and 1 will solve the second problem jointly by choosing level q_2^* which is the solution of the equation

$$\frac{\partial C_2}{\partial q_2} + \frac{\partial D_{22}}{\partial q_2} + \frac{\partial D_{12}}{\partial q_2} = 0 \qquad (17)$$

In practice both countries will first choose levels $\bar{q}_1$ and $\bar{q}_2$ and will then investigate together whether joint action would not enable them to improve their respective situations. In their negotiations country 1 might consider that it bore an "equivalent waste treatment cost" equal to $C_1 + D_{11}$ and imposed a damage cost of D_{21}, while country 2 might consider that it bore an "equivalent waste treatment cost" equal to $C_2 + D_{22}$ and imposed a damage cost of D_{12}, since the functions $(C_1 + D_{11})$ and $(C_2 + D_{22})$ have the usual properties of waste treatment cost functions when $q_1 < \bar{q}_1$ and $q_2 < \bar{q}_2$,

$$\frac{\partial (C_1 + D_{11})}{\partial q_1} < 0 \; ; \qquad \frac{\partial (C_2 + D_{22})}{\partial q_2} < 0$$

As it is probably only the lowering of levels $\bar{q}_1$ and $\bar{q}_2$ to the joint optimum levels q_1^* and q_2^* which will cause difficulties in international relations, the problem of reciprocal transfrontier pollution amounts in fact to two simultaneous problems of one-way pollution in which only an upstream country which suffers no damage is able to reduce the level of pollution in the downstream country.

As many cases of transfrontier pollution are inherently one-way (pollution of a river or coastal current) or are virtually one-way (air pollution in areas with prevailing winds), and as, moreover, an important group of reciprocal pollution problems consists of a combination of two cases of one-way pollution, it would appear to be a natural step to examine in detail the problem of one-way pollution.

STUDY OF AN EXTREME CASE OF ONE-WAY POLLUTION

An extreme case of one-way pollution is found when the cost of waste treatment $C(q)$ arises in the upstream country only and the damage cost $D(q)$ arises in the downstream country only. Let us assume that the damage cost is increasing, that the cost of waste treatment is decreasing and that the overall cost $T = C + D$ admits of only one minimum value when $q = q^*$. Let us also assume that the two countries are co-operative, i.e. that the upstream country is prepared to treat its wastes beyond the required level if the downstream country pays the cost of this additional treatment and that the downstream country will accept a higher pollution level than planned for if it is paid the additional damage cost.

The problem is how to find a formula for sharing the overall cost $T(q) = C(q) + D(q)$ between the two countries. The extreme solutions are for the upstream country to bear the overall cost T (civil liability principle) or for the downstream country to bear the overall cost T (victim-pays principle). In either case the upstream country will agree to reduce its discharges to the level q^* which minimises T. In the former case it will bear the cost of this operation, i.e. $C(q^*)$, and will pay the downstream country the value of the damage $D(q^*)$, while in the latter case it will be paid the cost $C(q^*)$ by the downstream country, which will in addition bear damage costs amounting to $D(q^*)$.

Uniform cost-sharing rule

We shall assume that both countries are prepared to adopt a uniform rule for sharing the overall cost which will give results which go part of the way towards those given by the civil liability principle or the victim-pays principle.

We shall examine below the broad category of linear formulas having three parameters a, b and E, in which the upstream country undertakes to bear the cost

$$T_m = aC(q) + (1-b) D(q) + E \qquad (18)$$

and the downstream country undertakes to bear the cost

$$T_v = (1-a) \, C(q) + bD(q) - E \tag{19}$$

This category includes all the formulas so far proposed for solving transfrontier pollution problems, viz:

a) The civil liability principle (a=1, b=0, E=0)

b) The victim-pays principle (a=0, b=1, E=0)

c) The principle of shared responsibility (a+b=1, E=0)

d) The principle of shared waste treatment costs (b=1, E=0)

as well as the polluter – pays principle as defined by the OECD, viz. $E = (1-a) \, C(q_c) - (1-b) \, D(q_c)$. If the two countries agree on a tolerable discharge levels q_c, the upstream country will bear the cost

$$T_m = a \, \underline{/}C(q) - C(q_c) \underline{/} + (1-b) \, \underline{/}D(q) - D(q_c) \underline{/} + C(q_c) \tag{20}$$

i.e. the cost of the pollution control measures $C(q_c)$ when the level of discharge is the agreed level q_c. Likewise, the downstream country will bear the cost

$$T_v = (1-a) \, \underline{/}C(q) - C(q_c) \underline{/} + b \, \underline{/}D(q) - D(q_c) \underline{/} + D(q_c) \tag{21}$$

i.e. the cost of the residual damage $D(q_c)$ when the level of discharge is the agreed level q_c. So we see that three parameters are required to cover all the principles so far proposed.

If the two countries agreed to adopt a three-parameter formula, the problem of transfrontier pollution would consist in choosing values for these three parameters which would satisfy the principles of economic efficiency and equity and would also reduce international tension.

In this paper we shall not broach the problem of equity, so that we shall only deal with choosing two of the parameters, e.g. a and b. We shall then have to choose formulas which satisfy the following criteria:

a) The cost borne by each country (T_m or T_v) shall not be higher than T = C + D or lower than zero.

b) The cost borne by each country shall have no analytical maximum and shall have a minimum not too far removed from the minimum value of T.

c) The optimum level q^* shall always be reached and each country shall bear a fraction of the cost $T(q^*)$.

d) The country which is polluted beyond the agreed level shall be indemnified for the extra cost it has to bear as compared with the agreed level.

Conditions (a) and (b) will be satisfied for $q < q_c$ if

$$0 \leqslant a \leqslant 1$$
$$0 \leqslant b \leqslant 1$$

and if E is not too large. Defining a uniform formula will consist in choosing a point inside the square in Figure 2.

Condition (d) means that the formula valid for $q > q_c$ will be

$$\left[\begin{array}{l} T_m = \underline{/}C(q) - C(q_c) \underline{/} + \underline{/}D(q) - D(q_c) \underline{/} + E + aC(q_c) + (1-b)D(q_c) \end{array} \right. \tag{22}$$
$$\left. T_v = (1-a) \, C(q_c) + bD(q_c) - E \right. \tag{23}$$

In this case the values chosen will be a = 1 and b = 0.

Figure 2

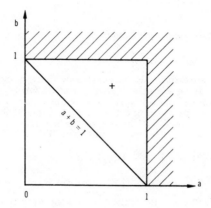

The polluter-pays principle, victim-pays principle and civil liability principle

If the cost-sharing formula has to be the same for levels both above and below the optimum level, the parameters a and b will be fixed (a = 1, b = 0). The basis for cost-sharing will then be the <u>civil liability principle</u> modified by a transfer payment having the effect of reducing the cost borne by the upstream country when pollution is at the optimum level:

$$\begin{cases} T_m = C(q) + D(q) - \alpha \left[C(q^\star) + D(q^\star) \right] = (T-T^\star) + (1-\alpha) T^\star & (24) \\ T_v = \alpha \left[C(q^\star) + D(q^\star) \right] = \alpha T^\star & (25) \end{cases}$$

with $0 < \alpha < 1$. If both countries adopted this principle, the only thing left for them to negotiate would be the value of the co-efficient α (problem of equity).

If, however, the cost-sharing formula is not to be the same when pollution is above the agreed level as when it is below it, parameters a, b and E will be determined by other considerations.

As the countries are assumed to have freedom of choice in defining their responsibility for transfrontier pollution, it would seem preferable for them to choose a formula which exactly corresponds to the facts and calls for no bargaining. It can be shown mathematically that this condition means that the value of q which minimizes T, T_m and T_v are the same, i.e. that a + b = 1.

When the two countries have difficulty in agreeing on the cost of the pollution control measures, but can agree on the damage cost, it is better for the downstream country not to bear any part of the cost of pollution control (a = 1), so as not to feel any need to challenge the estimate given by the upstream country. The latter country, in the belief that it knows the cost of pollution control, will then make the most rational choice of waste treatment level if it has to bear the cost

$$T_m = C + D + E \tag{26}$$

In this case, b = 0 and the downstream country will bear the cost

$$T_v = \quad - E \tag{27}$$

whatever the level of pollution. Parameter E will lie between $- \left[C(q^\star) + D(q^\star) \right]$ and 0, and will be a matter for negotiation between the two countries. This constitutes the principle of <u>civil liability</u> modified by the transfer payment E. In this case the downstream country will obtain recognition of its right to an environment which is not polluted beyond the optimum level in return for making a transfer payment to the upstream country.

When the two countries have difficulty in agreeing on an estimate of the damage cost, but are well informed as to the cost of pollution control, it will be better for the downstream country to bear alone the damage cost (b=1), which it thinks it can estimate. The latter country will make the most rational choice of pollution level if it has to bear the cost

$$T_v = C + D - E \tag{28}$$

In this case a $= 0$ and the upstream country will bear the cost

$$T_m = E \tag{29}$$

whatever the level of pollution. Parameter E will lie between 0 and $\overline{[C(q^*) + D(q^*)]}$ and will be a matter for negotiation between the two countries. In this case the upstream country will acquire the right to pollute up to the optimum level q^* in return for making a transfer payment E to the downstream country. The principle used here will be called the <u>modified victim-pays principle</u>. When $q > q_c$, recourse must be had to the formula

$$\begin{cases} T_m = C(q) - C(q_c) + D(q) - D(q_c) + E \tag{30} \\ T_v = C(q_c) + D(q_c) - E \tag{31} \end{cases}$$

The modified civil liability principle and the modified victim-pays principle can be used in conjunction with the <u>polluter-pays principle</u> (E = (1-a) $C(q_c)$ - (1-b) $D(q_c)$). In this case agreeing on a level q_c beyond which the upstream country will be responsible for any additional damage will mean determining the costs to be borne by each country when pollution is at the optimum level. The modified civil liability principle combined with the polluter-pays principle may be expressed thus:

$$\begin{cases} T_m = C(q) + D(q) - D(q_c) \tag{32} \\ T_v = D(q_c) \tag{33} \end{cases}$$

and the modified victim-pays principle combined with the polluter-pays principle may be expressed thus:

$$q < q_c \qquad \begin{cases} T_m = C(q_c) \tag{34} \\ T_v = C(q) - C(q_c) + D(q) \tag{35} \end{cases}$$

$$q > q_c \qquad \begin{cases} T_m = C(q) + D(q) - D(q_c) \tag{36} \\ T_v = D(q_c) \tag{37} \end{cases}$$

When the modified civil liability principle is adopted and level q_c is fixed, the cost borne by the downstream country will be independent of the amount of polluting activity in the upstream country and of developments in pollution control techniques.

When the modified victim-pays principle is adopted and level q_c is fixed, the cost borne by the upstream country will be independent of the amount of activity which suffers pollution in the downstream country and of the state of knowledge regarding the effects produced by pollutants. Moreover, the cost borne by the downstream country will not vary when the cost of waste treatment is a linear function of the quantity of pollutant produced: $q_m \overline{[C(q) = a(q_m - q)]}$. The modified victim-pays principle will then be preferred to the modified civil liability principle to the extent that it is roughly in line with the principle that the downstream country should not have to bear an increase in cost due to changes occurring in the upstream country, and to the extent that it agrees perfectly with the principle that the upstream country should not bear an increase in cost due to changes occurring in the downstream country.

Lastly, let us examine a case in which parameters a and b do not satisfy the condition a + b = 1. Let us assume that the values chosen are a $= 1$ and E $= 0$. The costs borne by countries 1 and 2 would then be C(q) and D(q) when $q < q_c$ and would be C(q) + D(q) - D(q_c) and $D(q_c)$ when $q > q_c$. Apparently the cost borne by the upstream country when $q < q_c$ would be independent of D(q) and cost borne by the downstream country would be independent of C(q). However, this formula does not represent the facts, because, if the upstream country chooses to treat its wastes up to the level q_c and so minimises the cost it has to bear, the downstream country will have to bear the residual damage cost $D(q_c)$ or the lesser cost $C(q^{\pm}) + D(q^{\pm}) - C(q_c) < D(q_c)$. The downstrean country will prefer to make a transfer payment of $C(q^{\pm}) - C(q_c)$ to the upstream country in return for more waste treatment by the latter. Thus both countries will support costs in accordance with the modified victim-pays principle, so that there seems to be no need to deal specially with this case.

The "gain" $C(q_c) + D(q_c) - C(q^{\pm}) - D(q^{\pm})$ will benefit only the downstream country, but may drop to nil if the optimum level $q^{\pm}$ becomes equal to q_c (e. g. if the cost of waste treatment increases). As the upstream country, on the same assumptions, will have to bear a greater increase in waste treatment costs, it is unlikely that a reduction in the "gain" would be a sufficient argument for not adopting the modified victim-pays principle.

SEPARABLE RECIPROCAL POLLUTION

The formulas for sharing the total cost when both countries have agreed on levels q_{1c} and q_{2c} for their pollution are obtained by combining the results of the analysis of two cases of one-way pollution. They are:

Modified civil liability principle

$$\left\{ \begin{aligned} &T_1 = \sqrt{C_1(q_1)} + D_{11}(q_1) + D_{21}(q_1) - D_{21}(q_{1c})\sqrt{7} + \sqrt{D_{12}(q_{2c})\sqrt{7}} && (38)\\ &T_2 = \sqrt{D_{21}(q_{1c})\sqrt{7}} + \sqrt{C_2(q_2)} + D_{22}(q_2) + D_{12}(q_2) - D_{12}(q_{2c})\sqrt{7} && (39) \end{aligned} \right.$$

Modified victim-pays principle

$$\begin{aligned} q_1 < q_{1c} \quad &\left\{ \begin{aligned} &T_1 = \sqrt{C_1(q_{1c})} + D_{11}(q_{1c})\sqrt{7} + \sqrt{C_2(q_2)} - C_2(q_{2c})\\ &\quad + D_{22}(q_2) - D_{22}(q_{2c}) + D_{12}(q_2)\sqrt{7} && (40)\\ q_2 < q_{2c} \quad &T_2 = \sqrt{C_1(q_1)} - C_1(q_{1c}) + D_{11}(q_1) - D_{11}(q_{1c}) + D_{21}(q_1)\sqrt{7}\\ &\quad + \sqrt{C_2(q_{2c})} + D_{22}(q_{2c})\sqrt{7} && (41) \end{aligned} \right. \end{aligned}$$

In both cases, when the discharges are at the agreed levels q_{1c} and q_{2c}, country 1 will bear the cost of waste treatment up to the agreed level $\sqrt{C_1(q_{1c})\sqrt{7}}$, plus the damage cost in country 1 due to discharges at agreed level q_{1c} in country 1 $\sqrt{D_{11}(q_{1c})\sqrt{7}}$, plus the residual damage cost in country 1 due to discharges at agreed level q_{2c} in country 2 $\sqrt{D_{12}(q_{2c})\sqrt{7}}$. Similarly country 2 will bear the cost

$$C_2(q_{2c}) + D_{22}(q_{2c}) + D_{21}(q_{1c}).$$

If both countries adopt the modified civil liability principle, country 1 will bring its discharges down from agreed level q_{1c} to the optimum level $q_1^{\pm}$ and country 2 will bring its discharges down from agreed level q_{2c} to the optimum level $q_2^{\pm}$. If both countries adopt the modified victim-pays principle, country 1 will ask country 2 to reduce its discharges from agreed level q_{2c} to the optimum level $q_2^{\pm}$ and country 2 will ask country 1 to reduce its discharges from agreed level q_{1c} to the optimum level $q_1^{\pm}$. If both countries agree in their estimates of reciprocal damage D_{12} and D_{21}, but not necessarily on the waste treatment costs and damage costs which they incur as a result of their own discharges, they can easily arrive at the economic optimum by adopting the modified civil liability principle. If, on the other hand, they agree in their estimates of the cost of pollution control and the damage cost which they incur as a result of their own discharges, they can easily arrive at the economic optimum by adopting the modified victim-pays principle, even if they fail to agree on the reciprocal damage cost.

In many cases the agreed levels will in fact be the levels of discharge which are actually observed when there is no joint policy for managing the shared environment. In other words, q_{1c} and q_{2c} will be the levels which minimise the sum of the waste treatment cost in each country and the damage cost in that same country (individual optimum). Level q_{1c} will minimise $(C_1 + D_{11})$ and level q_{2c} will minimise $(C_2 + D_{22})$.

In these conditions, if the two countries agree in their estimates of waste treatment costs C_1 and C_2, they will estimate their marginal damage costs at

$$\left[\frac{\partial D_{11}}{\partial q_1}\right]_{q_{1c}} \quad \text{and} \quad \left[\frac{\partial D_{22}}{\partial q_2}\right]_{q_{2c}}$$

on the basis of which they can estimate roughly the additional damage costs $D_{11}(q_1) - D_{11}(q_{1c})$ and $D_{22}(q_2) - D_{22}(q_{2c})$. On this assumption the two countries would only have to agree in estimating waste treatment costs $C_1(q_1)$ and $C_2(q_2)$ in order to be able to arrive at the economic optimum.

Comments

1. If pollution is one-way ($D_{12} = 0$), the level q_2 which corresponds to the individual optimum will be the same as the level which corresponds to the joint optimum. Upstream country 2 will choose this value for q_2 and the only problem will be how to determine the discharge level q_1 in country 1 (downstream).

2. If the two countries select levels q_{1c} and q_{2c} corresponding to the individual optimum, they will bring down the costs T_1 and T_2, which they have to bear, by reducing their discharges to the optimum levels q_1^* and q_2^*. This does not necessarily mean that the sum of the waste treatment cost and damage cost in one country will be less for the optimum discharges than for the agreed discharges, since the effect of adopting a cost-sharing formula (T_1 and T_2) will be to alter the cost borne by each country ($T_1 \neq C_1 + D_1$ and $T_2 \neq C_2 + D_2$).

NON-SEPARABLE RECIPROCAL POLLUTION

When reciprocal pollution is not separable, no cost-sharing formula can be derived from the elementary case of one-way pollution, because non-separability introduces an additional cost, due to interaction (overloading) between polluters which cannot easily be debited to either country. We shall now examine two formulas embodying the modified victim-pays principle.

If two necessary conditions are that the cost-sharing formula reduces to the one obtained above for the case when pollution is separable and that the amount to be debited to a country for the interaction effect on the environment is calculated from the damage cost in that country only, the following is a cost-sharing formula which is in line with the modified victim-pays principle:

$$
\begin{aligned}
T_1 = & \: /\!\!\!\!\:C_1(q_{1c}) + D_1(q_{1c}, q_{2c})\!\!\:/ + /\!\!\!\!\:C_2(q_2) - C_2(q_{2c}) + D_2(q_{1c}, q_2) - D_2(q_{1c}, q_{2c}) \\
& + D_1(q_{1c}, q_2) - D_1(q_{1c}, q_{2c})\!\!\:/ \\
& + /\!\!\!\!\:D_1(q_1, q_2) - D_1(q_{1c}, q_2) + D_1(q_{1c}, q_{2c}) - D_1(q_1, q_{2c})\!\!\:/ \qquad (42)
\end{aligned}
$$

$$
\begin{aligned}
T_2 = & \: /\!\!\!\!\:C_2(q_{2c}) + D_2(q_{1c}, q_{2c})\!\!\:/ + /\!\!\!\!\:C_1(q_1) - C_1(q_{1c}) + D_1(q_1, q_{2c}) - D_1(q_{1c}, q_{2c}) \\
& + D_2(q_1, q_{2c}) - D_2(q_{1c}, q_{2c})\!\!\:/ \\
& + /\!\!\!\!\:D_2(q_1, q_2) - D_2(q_1, q_{2c}) + D_2(q_{1c}, q_{2c}) - D_2(q_{1c}, q_2)\!\!\:/ \qquad (43)
\end{aligned}
$$

The third term of each formula corresponds to the "overloading" cost and it will be nil if

$$\frac{\partial^2 D_i}{\partial q_1 \, \partial q_2} = 0$$

If countries 1 and 2 choose levels q_{1c} and q_{2c} corresponding to the individual optimum and then adopt the above cost-sharing formula country 1 will request country 2 to reduce its discharges to the level given by $\dfrac{\partial T_1}{\partial q_2} = 0$, i.e.

$$\frac{\partial C_2(q_2)}{\partial q_2} + \frac{\partial D_2(q_1, q_{2c})}{\partial q_2} + \frac{\partial D_1(q_1, q_2)}{\partial q_2} = 0 \qquad (44)$$

and country 2 will request country 1 to reduce its discharges to the level given by

$$\frac{\partial T_2}{\partial q_1} + \frac{\partial C_1(q_1)}{\partial q_1} + \frac{\partial D_1(q_1, q_{2c})}{\partial q_1} + \frac{\partial D_2(q_1, q_2)}{\partial q_1} = 0 \qquad (45)$$

It can easily be shown that the two countries will successively choose levels approaching the optimum solution, but they will never attain it by taking independent decisions. The above formula cannot, therefore, lead to economic efficiency through independent decisions taken by the two countries.

This is the result of the particular choice of "overloading" cost introduced into the cost-sharing formula.

We now suggest a formula which does lead to the economic optimum while still complying both with the modified victim-pays principle and with the polluter-pays principle. The costs borne by the two countries will be:

$$T_1 = \lfloor C_1(q_{1c}) + D_1(q_{1c}, q_{2c}) \rfloor + \lfloor C_2(q_2) - C_2(q_{2c}) + D_1(q_{1x}, q_2)$$
$$- D_1(q_{1x}, q_{2c}) + D_2(q_{1x}, q_2) - D_2(q_{1x}, q_{2c}) \rfloor$$
$$+ \lfloor D_1(q_1, q_2) - D_1(q_{1c}, q_{2c}) + D_1(q_{1c}, q_2^*) - D_1(q_1, q_2^*)$$
$$+ D_1(q_{1x}, q_{2c}) - D_1(q_{1x}, q_2) \rfloor \qquad (46)$$

$$T_2 = \lfloor C_2(q_{2c}) + D_2(q_{1c}, q_{2c}) \rfloor + \lfloor C_1(q_1) - C_1(q_{1c}) + D_2(q_1, q_{2x})$$
$$- D_2(q_{1c}, q_{2x}) + D_1(q_1, q_{2x}) - D_1(q_{1c}, q_{2x}) \rfloor$$
$$+ \lfloor D_2(q_1, q_2) - D_2(q_{1c}, q_{2c}) + D_2(q_1^*, q_{2c}) - D_2(q_1^*, q_2)$$
$$+ D_2(q_{1c}, q_{2x}) - D_2(q_1, q_{2x}) \rfloor \qquad (47)$$

where q_{1x} and q_{2x} are the discharge levels chosen successively by the two countries and q_1^* and q_2^* are the discharge levels at the economic optimum (or the asymptotic values of q_{1x} and q_{2x} if the countries do not choose levels q_1^* and q_2^*). When $q_1 = q_{1c}$ and $q_2 = q_{2c}$, the countries will bear the cost given by the first term between square brackets. When $q_1^* = \lim_{t \to \infty} q_{1x}$ and $q_2^* = \lim_{t \to \infty} q_{2x}$, they will bear the costs

$$T_1^* = C_1(q_{1c}) + \lfloor C_2(q_2^*) - C_2(q_{2c}) \rfloor + D_2(q_1^*, q_2^*) - D_2(q_1^*, q_{2c})$$
$$+ D_1(q_{1c}, q_2^*)$$

$$T_2^* = C_2(q_{2c}) + \lfloor C_1(q_1^*) - C_1(q_{1c}) \rfloor + D_1(q_1^*, q_2^*) - D(q_{1c}, q_2^*)$$
$$+ D_2(q_1^*, q_{2c})$$

The sum of the costs borne by the two countries will be equal to $T = C + D$ if they take q_{1x} and q_{2x} as estimates for q_1^* and q_2^*. The total differential of $T_1(q_1, q_2)$ will be

$$dT_1 = \left\lfloor \frac{\partial C_2(q_2)}{\partial q_2} + \frac{\partial D_2(q_{1x}, q_2)}{\partial q_2} + \frac{\partial D_1(q_1, q_2)}{\partial q_2} \right\rfloor dq_2 +$$
$$+ \left\lfloor \frac{\partial D_1(q_1, q_2)}{\partial q_1} - \frac{\partial D_1(q_1, q_2^*)}{\partial q_1} \right\rfloor dq_1$$

Country 1 will therefore choose a level q_2 to satisfy

$$\frac{\partial C_2(q_2)}{\partial q_2} + \frac{\partial D_2(q_{1x}, q_2)}{\partial q_2} + \frac{\partial D_1(q_1, q_2)}{\partial q_2} = 0 \qquad (48)$$

assuming that $q_1 = q_{1x}$. It will not choose to alter level q_1, as the marginal cost

$$\frac{\partial D_1(q_1, q_2)}{\partial q_1} - \frac{\partial D_1(q_1, q_2^*)}{\partial q_1} \approx \frac{\partial^2 D_1(q_{1x}, q_{2x})}{\partial q_1 \; \partial q_2} (q_2 - q_{2x})$$

will usually be small in relation to the marginal cost of waste treatment $\dfrac{\partial C_1}{\partial q_1}$ which it would have to bear it if decided to alter q_1 unilaterally.

Similarly, country 2 will choose a level q_1 which satisfies

$$\frac{\partial C_1(q_1)}{\partial q_1} + \frac{\partial D_1(q_1, q_{2x})}{\partial q_1} + \frac{\partial D_2(q_1, q_2)}{\partial q_1} = 0 \tag{49}$$

assuming that $q_2 = q_{2x}$.

The choices of q_1 and q_2 made by countries 2 and 1 will be modified in accordance with the successive values of q_{2x} and q_{1x}.

This iterative process will generally converge towards the optimum solution q_1^* and q_2^* when function T is strictly convex in the area of the variables being considered

$$\text{(i.e. if } \quad \frac{\partial^2 T}{\partial q_1^2} \cdot \frac{\partial^2 T}{\partial q_2^2} > \left[\frac{\partial^2 T}{\partial q_1 \; \partial q_2}\right]^2 \text{)} \tag{50}$$

The cost-sharing formulas (46) and (47) will be usable if the two countries agree in their estimates of waste treatment costs C_1 and C_2 and of the additional damage costs

$$D_1(q_1, q_{2x}) - D_1(q_{1c}, q_{2x}) \approx \frac{\partial D_1}{\partial q_1} (q_1 - q_{1c}) \text{ and } D_2(q_{1x}, q_2) - D_2(q_{1x}, q_{2c}) \approx$$

$$\frac{\partial D_2}{\partial q_2} (q_2 - q_{2c})$$

but they require no agreement on damage costs D_1 and D_2 or on marginal reciprocal damage costs

$$\frac{\partial D_1}{\partial q_2} \text{ and } \frac{\partial D_2}{\partial q_1} \quad .$$

These formulas will therefore make it possible to solve problems of non-separable transfrontier pollution when the modified victim-pays principle meets the wishes of the two countries.

RECIPROCAL TRANSFRONTIER POLLUTION WITH POLLUTION TARGETS

The transfrontier pollution problem here is the same as in Annex 2, but assumes that the two countries are wholly unable to determine the damage cost. It is assumed that country 1 will tolerate no increase beyond P_1 in the level of pollution $p_1(q_1, q_2)$ affecting it and that country 2 will not tolerate a pollution level $p_2(q_1, q_2)$ in excess of P_2.

Generally speaking, the conditions
$$\begin{cases} p_1(q_1, q_2) = P_1 \\ p_2(q_1, q_2) = P_2 \end{cases}$$

can be satisfied simultaneously. Let us start by examining the case illustrated in Figure 1. Point D corresponds to discharges which satisfy the targets of the two countries, while OADC is the permitted area for discharges. Let $C = C_1(q_1) + C_2(q_2)$ = constant be the equation of the locus of the points corresponding to a given total waste treatment cost (isocost), and B be the point of tangency of the isocost with the contour of permitted area OADC. Point B will sometimes coincide with point D, so that there will be contact but not tangency. The "optimum" levels of discharge q_1^* and q_2^* are shown by point B.

The initial conditions lie along or inside the contour of the area ABDCI, since countries 1 and 2 cannot tolerate discharges q_1 and q_2 in excess of those given by
$$\begin{cases} p_1(q_1, 0) = P_1 \\ p_2(0, q_2) = P_2 \end{cases}$$

The equity problem in transfrontier pollution is how to share the optimum waste treatment cost $C_1(q_1^*) + C_2(q_2^*)$ between the two countries. An allocation whereby country 1 would bear $C_1(q_1^*)$ and country 2 would bear $C_2(q_2^*)$ might appear wholly inequitable, particularly if these costs differed while all the other features of the problem were identical.

Figure 2 shows the case in which there are no discharge levels corresponding to the targets of countries 1 and 2. Thus, the amount of waste treatment required to satisfy both countries will be greater than one of them would consider necessary. If points H and B are the tangency points at which the targets of countries 1 and 2 meet the isocosts, the two countries will have to share the waste treatment cost C(H) starting from an initial situation lying inside AICC'BA, and country 2 will doubtless be obliged to bear the further treatment cost C(B) - C(H) (or only part of it if country 1 agrees to assist country 2, either on grounds of equity or in order to improve its own environment). The "optimum" discharges will be the q_1^* and q_2^* co-ordinates of point B.

Figure 3 illustrates the situation when pollution is one-way (country 1 is not affected by omissions from country 2). Country 1 may take the view that it should not have to bear additional treatment costs to bring its discharges down to point B (instead of H). Country 2 will stress the environmental improvement in country 1 when discharges are at q_1^* and q_2^* and will advance the concept of "equitable" sharing between the two countries, of the assimilative capacity of the environment in country 2.

When treatment costs $C_1(q_1)$ and $C_2(q_2)$ are known and the optimum point B can be determined, equivalent damage functions can be constructed:
$$\begin{cases} D_1(p_1) - D_1(P_1) = \alpha \left[a_{11}(q_1 - q_1^*) + a_{12}(q_2 - q_2^*) \right] \\ D_2(p_2) - D_2(P_2) = \beta \left[a_{21}(q_1 - q_1^*) + a_{22}(q_2 - q_2^*) \right] \end{cases}$$

Figure 1

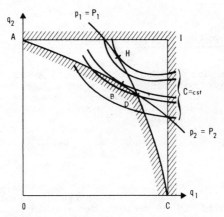

Figure 2

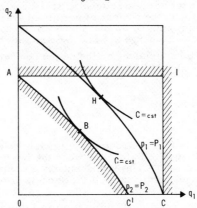

Figure 3

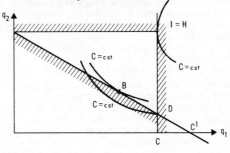

where D_1 and D_2 are the damage costs in countries 1 and 2, so that the minimum overall cost $T = C_1+C_2+D_1+D_2$, is situated exactly at point B. The four parameters $\alpha\, a_{11}$, $\alpha\, a_{12}$, $\beta\, a_{21}$, and $\beta\, a_{22}$ can be determined from the two marginal treatment costs at point B and from the two tangents to the isopollution curves at point B, thus:

$$
\begin{cases}
\dfrac{\partial p_1 \,/\, \partial q_2}{\partial p_1 \quad \partial q_1} = \dfrac{a_{12}}{a_{11}} \\[3ex]
\dfrac{\partial p_2 \,/\, \partial q_2}{\partial p_2 \quad \partial q_1} = \dfrac{a_{22}}{a_{21}}
\end{cases}
$$

This very approximate information on damage costs enables us to deal with the problem in the same way as the one before and to define the responsibility of each country when discharges are not at the optimum level. If this method cannot be accepted, it would appear to be extremely difficult to measure the economic consequences of diverging from the optimum, or to suggest any formula for sharing the extra cost $T - T^*$ due to such divergence.

THE MUTUAL COMPENSATION PRINCIPLE: AN ECONOMIC INSTRUMENT FOR SOLVING CERTAIN TRANSFRONTIER POLLUTION PROBLEMS

Note by the Secretariat

INTRODUCTION

The leading economic instruments used for solving national pollution problems are taxes and charges levied against polluters, charges paid by those benefiting from treatment schemes (water for drinking or irrigation), and incentives ("bribes") to polluters for reducing pollution. Making separate use of these instruments means introducing a single charge or tax to correct externalities caused by pollution. They can be effective if the taxed transactors seek to minimise the costs due to pollution.

Another possibility yet to be availed of is to create marketable "pollution certificates" (1). The polluters would buy these certificates from the pollutees whenever they wanted to increase the level of pollution while the pollutees would purchase them from the polluters in order to achieve a less polluted environment. The price of the certificate would depend on the utility of a pollution unit for the polluters and the disutility to the pollutee of the same unit. Under the scheme two charges would hence be used for correcting externalities.

Since economic instruments imposed by the public authorities have so far alone been used for simulating a market at national scale, it may be well to study a system much like the certificate scheme but not dependent on the establishment of a "certificate" market.

This report deals with the study of an economic instrument applicable to one-way transfrontier pollution which increases mutual trust and leads to economic efficiency. This instrument is based on the symmetrical principle of "taxing" the two countries or, in other terms, on the simultaneous sale of pollutant "emission" and "treatment" rights by an agency holding such rights. This instrument comprises lump sum transfers offered by the agency to the countries in order that each country's net cost should be the one which corresponds to the agreed equity principle or the agreed initial rights.

This study only covers the case of countries with a benevolent and friendly attitude and comparable standards of living. In this situation it may reasonably be assumed that the marginal utility of a given expenditure in each country is the same and that overall welfare is independent of how the costs due to transfrontier pollution are shared. In particular, the welfare of a country is not affected by the absolute or relative amount of the costs due to transfrontier pollution borne by the other country but only by the amount of these costs that it bears itself.

The cost of goods and services in a country does not depend to any significant extent on what fraction of costs due to transfrontier pollution is borne by the other country since these costs are generally small compared with the national product and because trade between the two countries is governed by world prices (for example: if the polluting country has to undertake treatment, the polluted country does not, in the end, pay the cost of this treatment, unlike the national situation where the community generally ends up by paying the cost of treatment operations carried out by the polluters). With these assumptions, cost-benefit analysis can be invoked and problems of efficiency can be handled separately from those of equity. These assumptions imply that the optimal pollution level does not depend on how much of the cost due to transfrontier pollution is borne by each country. Similarly, neither the damage nor the treatment cost depends on the equity principle the countries adopt for sharing transfrontier pollution costs since the costs are generally insignificant compared with the national product.

Because of the assumption of benevolence, it is taken for granted that the countries will not try to reduce the welfare of another country while maintaining their own at the same level (hostility); they will not oppose an increase in the welfare of another country while their own remains constant, and will not insist in

1) This idea was introduced by Dales and later taken up by several authors. It is considered in detail in Annex 2. It is introduced in France in matters of building permits.

sharing in the increase in another country's welfare on the pretext of maintaining their relative situations (envy, jealousy). A generalisation of this study to cover the case of malevolent countries is given in Annex 4 where it is shown that the mutual compensation principle is easily extended if the degree of malevolence is known at the outset.

Finally, for reasons of simplicity the paper does not go into the cost of information, negotiation and settlement even though these costs can play some part in the choice of the optimal level and can depend on the equity principle adopted.

Compared with national pollution, the problems of transfrontier pollution are marked by three main differences:

a) the groups involved (regions, countries) are few in number, organised, and capable of negotiating;

b) the groups involved have no common body set up for maximising general well-being;

c) there is no supranational authority for resolving problems of transfrontier pollution, nor any machinery for redistributing wealth between the countries.

BACKGROUND

In a previous study (see pages 87-114) it was shown that transfrontier pollution could be dealt with by resorting to a single tax (1), whether applied to the polluting country when damage costs in the polluted country were known and accepted by both countries, or to the polluted country when the cost of control in the polluting country was known and accepted by both countries.

---

1) The principle of taxing the polluter is being applied at national level under the schemes of incentive charges imposed on the polluter. These charges are calculated on the basis of damage cost, whether it is estimated by experts or determined by the polluted party through such decision-making procedures as the setting of quality objectives by a designated "board" or by an elected assembly representing all the parties involved. While the difficulties in calculating damage are of course great, they do not seem insurmountable, since the public authorities are equipped to take decisions on a national scale (perhaps after some trial and error owing to the dearth of information).

The principle of taxing the victims of pollution seldom receives nation-wide application because the victims are usually numerous, poorly informed and sometimes unresponsive to a tax designed to alter their behaviour. Taxes do however exist for the purpose of discouraging settlement in certain areas which are already polluted and others are imposed on water used for drinking and irrigation. If the tax is calculated in terms of the cost of any additional pollution control made necessary by the polluted party's decisions, in theory the victims estimate the damage they suffer as accurately as possible since they must weigh the advantage resulting from pollution abatement against the disadvantage of having to pay a heavier price to abate it. Moreover they will themselves introduce measures for reducing the damage and will come to believe that they cannot hope for any later change in the allocation of environmental rights (for example, the resiting of a polluting installation under the pressure of public opinion after the adjacent area has been urbanised without those benefiting from the resiting measure having to contribute directly towards the resiting costs or being taxed on the increased value of the land thus achieved).

Moreover, when the environmental rights are assigned to the polluters and the victim-pays principle is therefore the one which is applicable, the victims are in fact subjected to a tax equivalent to the cost of pollution control while the polluters receive a bribe equivalent to the tax paid by the victims in order that suitable measures selected by the victims can be implemented.

The effect of a single tax is to cause the country which pays the tax to choose the optimal level of pollution and to assess as exactly as possible the amount of the cost it bears in the first instance (the cost of damage or of pollution control), regarding which any agreement seemed difficult to reach.

The taxed country thus freely assesses the cost which is hardest to calculate and on which the countries have failed to agree. It will provide an estimate which is as accurate as possible, since it will derive no advantage from some other choice.

The single tax principle thus appears particularly suited as a solution to problems of one-way transfrontier pollution whenever one of the reported costs is accepted but the other might not exactly match the economic, social and political facts in the country which bears this cost in the first instance.

THE PRINCIPLE OF MUTUAL COMPENSATION

When pollution control costs and damage costs are both considered as values regarding whose estimation countries do not easily agree, the suggestion offered here is recourse to a system of mutual taxation (double tax) combining the above principles of taxing the polluter or of taxing the pollutee. Under this scheme the polluting country pays a "pollution tax" related to the cost of damage estimated by the polluted country while the polluted country pays a "treatment tax" related to the cost of pollution control as estimated in the polluting country. The purpose of the "pollution tax" is to induce the polluting country to take suitable pollution-control measures and the "treatment tax" is intended to encourage the polluted country to accept the cost of residual damage. The system can be used for degrees of pollution matching finite evaluations of costs on a collective basis. Like all economic instruments, it is based on the assumption that countries try to minimise the total costs (1) due to pollution which they are called upon to bear according to a pre-determined set of "game rules".

The term "mutual compensation principle" stems from the fact that the country requiring more intensive treatment must compensate by paying a higher treatment tax, whereas the country requiring a higher level of pollution compensates by paying a heavier pollution tax. Another aspect of compensation is that the tax paid by one country permits the other country to receive compensation (a bribe).

The mutual compensation principle calls for the establishment of a joint agency (2) for the two countries, whose duties will be to record cost estimates reported by the countries, to deduce the acceptable level of pollution or waste treatment by equating marginal costs, and to collect and transfer the taxes. The countries retain complete freedom of decision in estimating the costs they bear in the first instance but realise that it is in their own interest to provide the most accurate estimate of such costs (see below). Hence the acceptable pollution level corresponds to the optimum since the costs reported are true costs, or, alternatively, the best available estimate of real costs. Moreover, the pollution levels required by each country correspond to the optimum level, since the effect of the two taxes is to give each country the same cost function to minimise. If instead, the polluted country had no tax to pay it would find it worth while to require that the pollution level be close to zero so that the cost of residual damage would be as low as possible.

Figure 1 shows financial flows and information flows implementing the mutual compensation principle. Figures 2 and 3 indicate how the problem of minimising costs in different countries can be equated to two identical national problems of minimising the sum of a cost in the country which bears it initially and a tax paid by this country to the joint agency (3). The joint agency acts in conformity with the wishes of the two countries if it minimises the sum of the collected taxes.

1) The total costs include all the indirect effects such as unemployment or any fall in economic growth potential.

2) Systems with and without a joint agency are discused in more detail in Annex 2.

3) With the proposed system, both countries can achieve economic efficiency if this is their goal at national level. The system is of special value in international questions where no institutions exist for maximising collective welfare.

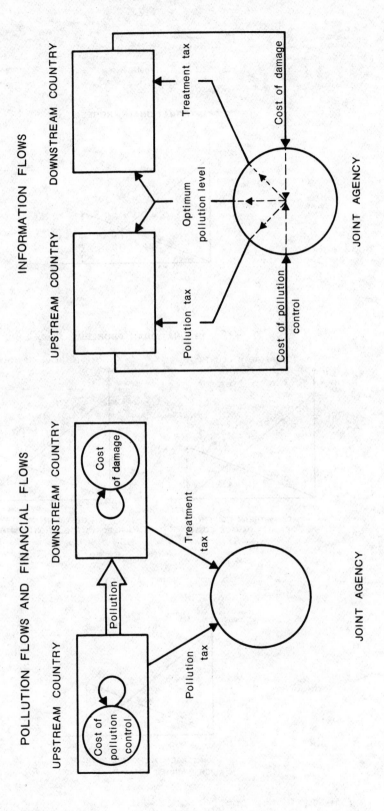

Figure 1

ONE-WAY TRANSFRONTIER POLLUTION

MUTUAL COMPENSATION PRINCIPLE

INFORMATION FLOWS

POLLUTION FLOWS AND FINANCIAL FLOWS

Figure 2

INTERNATIONAL PROBLEM

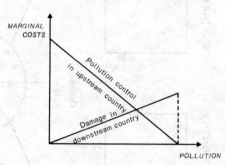

TWO NATIONAL PROBLEMS

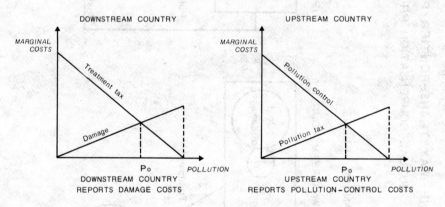

JOINT AGENCY

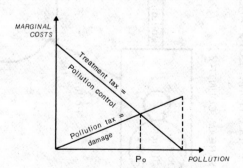

The system described has <u>three components</u> (two countries and a joint agency) and cannot be reduced to two components (two countries). A system based on the mutual compensation principle is therefore quite different from those considered in another report (see pp. 87-114) or systems based upon direct negotiations. In Annex 1 it is shown that the joint agency's budget cannot be strictly balanced. The "floating" character of the agency's budget is a peculiar difficulty of the system and is due to explicitly taking into account uncertainties in the matter of cost evaluation through time.

Figure 3

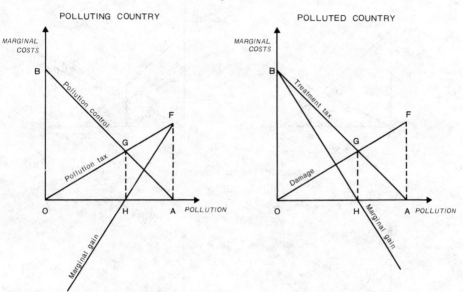

When the polluting country pays a pollution tax it maximises the total gain as measured from a non-tax situation (OA) by taking the tax into consideration and chooses treatment HA. When the polluted country pays a treatment tax it maximises the total gain as measured from a non-tax situation (zero pollution) by taking tax into consideration and chooses pollution level OH. Both the polluting and polluted countries therefore have the same goals corresponding to the overall optimum.

The fact that the countries invariably decide to announce the most accurate estimate of the costs they bear in the first instance is demonstrated in Figures 4 to 7. An analytic approach to the problem is described in Annex 1.

Assuming that the pollution country reports a pollution control cost in excess of its real value, the level of treatment selected by the joint agency will be below the optimal level. The polluting country will pay less for pollution control but a higher tax, since the pollution level will be above optimum. Figure 4 shows that the polluting country must bear a greater total cost if the cost of pollution has been overestimated. The converse case is illustrated in Figure 5, and also leads to the conclusion that the polluting country will find it advantageous to provide an exact estimate of the costs as best it can.

If the polluted country is assumed to report damage costs in excess of their real value, then the pollution level selected by the joint agency will be below the optimum level. While the polluted country will pay lower damage costs the treatment tax will be higher, since the level of treatment is above optimum. Figure 6 thus shows that the polluted country will bear a higher cost if the cost of damage is overestimated. The converse is dealt with in Figure 7, the conclusion here too being that the polluted country will find it advantageous to provide as accurate a value of damage cost as possible.

If the countries have no very accurate idea of the costs they initially bear, they will find that the system supplies an economic incentive for more exactly determining such costs in their own economic, social and political context, since

121

EFFECT OF INACCURATE COST ESTIMATES

HATCHED AREA : Cost borne by the country providing a wrong estimate, not taking the value of the constant into consideration.

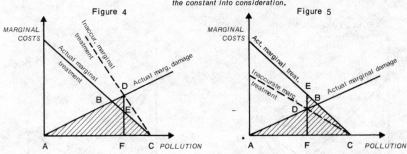

UPSTREAM COUNTRY

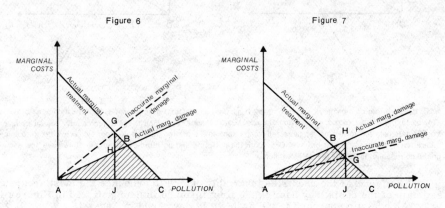

DOWNSTREAM COUNTRY

FIGURES 4 AND 5

The joint agency determines pollution level AF and treatment level FC. The upstream country bears treatment cost EFC and pollution tax **ADF**. Hence it bears an additional cost DEB if it provides a wrong estimate of the treatment cost. The downstream country bears a net cost ADC.

FIGURES 6 AND 7

The joint agency determines pollution level AJ and treatment level JC. The downstream country bears damage cost AHJ and treatment cost JGC. Hence it bears an additional cost GBH if it provides a wrong estimate of the damage cost. The upstream country bears a net cost AGC.

N.B. The existence of additional costs (DEB and GBH) does not depend on the assumption of linearly increasing marginal costs. The final result is not changed if the lines of marginal costs are curved.

they benefit from any improvement to their estimates. This information involves a cost which must not be too high in comparison with the optimisation gain. With the proposed system, the information cost is not incumbent upon a single country, but to each country as regards the cost it bears in the first instance and can estimate more easily.

COST ESTIMATES

In order to implement the mutual compensation principle the countries must agree to report, in monetary terms, the cost of pollution-control measures (upstream country) or damage costs (downstream country) for different pollution levels, or additional costs for variations around a pollution reference level. So long as such costs are known to the country which bears them in the first instance no difficulty should here arise (in practice, these costs are not well known and it is not possible to count on reaching the theoretical optimum corresponding to perfect information). Any comparison or conversion of these costs by the joint agency depends upon a jointly accepted exchange rate, and it will sometimes be necessary to make special arrangements to allow for variations in exchange rates during the period covered by an agreement.

Damage costs are hard to ascertain. While the downstream country might base its calculations on available scientific data it will have trouble in estimating the impact of certain applicable sociological, economic or political factors. It might in theory (1) define through the medium of its institutions what cost it is prepared to bear in order to reach an acceptable pollution level in relation to various hypothetical treatment taxes and thus ascertain how much it is prepared to pay, whereupon it would be able to deduce the cost of damage (Figure 8).

Figure 8

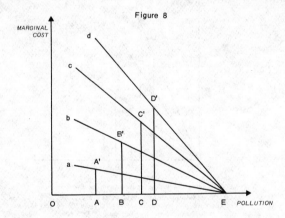

OE : Pollution level when no damage exists in the downstream country.

For different values of the treatment tax (rates a, b, c, d) the downstream country estimates that is prepared to pay AA'E for a pollution level under OA; BB'E for a level under OB; etc.

The marginal cost of damage is obtained by interpolation between O, A', B', C', D' and by extrapolation beyond these points.

This method should be applicable to the case under consideration, since a collective decision-making process comes into play which is followed by actual payment. It cannot be presumed to result in unrealistic evaluations, since if the institution of the polluted country reports a cost which fails to match the current economic, social and political facts, it knows that the downstream country will be penalised. The treatment tax thus emerges as the mechanism enabling the downstream community to state freely and effectively what value is attached to a largely unpolluted environment (the amount borne by the regional or national budget) and to ascertain what the cost of damage actually is. In the absence of such a tax the

1) See next page.

danger is that the cost of damage might be wrongly calculated, since the downstream country might often gain by overestimating this cost.

If the downstream country knows what the treatment tax is, it will only report the level of acceptable pollution and the corresponding marginal damage cost. It will however have to determine other values for marginal damage whenever the treatment tax is changed. In the event that the collective decision-making process only brings out one value for marginal damage cost, the authorities of the downstream country will have to calculate other values for it in the light of all available information.

This calculation will not affect the cost borne by the downstream country, but at least in theory it can substantially influence the cost borne by the upstream country. The hypothesis of smooth international relations (benevolent attitude) largely nullifies the hypothesis that the downstream country will wrongly estimate costs in order to injure the upstream country. In the extreme case, however, the latter might perhaps ask for the data used to calculate the shape of the marginal-damage curve between zero effect and the acceptable pollution level.

The cost of pollution control is not always a well-known factor, since besides the cost of certain specific measures (such as treatment) it also includes the cost of any human and technical under-employment which may follow the adoption of certain pollution standards. It coud also be evaluated in the upstream country by means of a collective decision-making process (Figure 9). The pollution tax is shown to be a mechanism whereby the upstream community can freely and effectively state the value attached to environmental pollution. It moreover has the effect of inducing the upstream country freely to choose the best possible strategy for abating pollution and to carry it through to the point where the marginal cost of abatement to the upstream community equals the rate of the pollution tax.

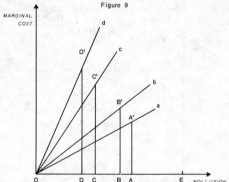

Figure 9

OE : Pollution level when no damage exists in the downstream country.

For different values of the pollution tax (rates a, b, c, d) the upstream country estimates that it is prepared to pay OAA' to avoid treating more than AE; OBB' to avoid treating more than BE; etc.

The marginal cost of pollution control is obtained by interpolation between E. A, B, C, D and by extrapolation beyond these points.

Note 1) of the page 123:

It will be noted that when part of the damage includes some loss of human life the polluted country may perhaps determine a tolerable pollution level (or risk level) in relation to a treatment tax but will be somewhat reluctant to associate any explicit monetary cost of damage with this level. The joint agency may however report a pollution tax to the polluting country which is not the price paid to the polluted country for human lives lost but is only an economic instrument for inducing the polluting country to take the damage caused into account. No doubt this will be impossible to use when a causal link between pollution and some specific loss of human life is established (e.g. poisoning following the local discharge of toxic substances). In this study pollution levels are assumed to be so low that loss of human life only occurs statistically i.e. such that no individual polluter can be held specifically liable. More generally, damage cost estimates are assumed to be finite within the area of pollution-level variation here considered.

Upon first implementing an agreement based on the mutual compensation principle the countries will no doubt provide cost estimates based on inadequate information. Any change over time in declared costs should not be interpreted as a strategic ploy for so adjusting the declared figures to those of the other country as to gain some specific advantage since the precise effect of the mutual compensation principle is that it obviates any such behaviour. Actually, it will mean that cost estimates are being improved owing to more complete and reliable information, thus promoting a change in the direction of increasing collective well-being (1).

It therefore appears that the system proposed can be used even when the data are imperfectly known or contain uncertainties. The countries need not comply with earlier estimates and will be able at any time to re-estimate costs in order to make allowance for new information, new discoveries or inventions or technological developments. Thus the mutual compensation principle appears as a method which enables the best solutions to be chosen in the light of current knowledge and attitudes (2). This result is obtained because the mutual compensation principle discourages any wrong determination of costs and hence tends to lead to the most efficient solution. Figure 10 illustrates the different phases in implementing an agreement when the countries have adopted an equity principle. Examples are discussed in Annex 3.

---

1) Changes with time

Even if it is possible to define an optimal pollution level at the time of signature of an agreement on transfrontier pollution, this level is unlikely to remain the same because of developments in pollution control techniques in the assessment of immediate or future damage, in the location and extent of polluting or pollution-sensitive activities and the exposed population, in relative price levels, exchange rates and so on. With the mutual compensation principle, it is possible to maintain pollution at its optimal level in spite of all these variations. In practice, the pollution level will only be changed at intervals which depend on the operating flexibility of available plant or the necessary lead time for new capital projects.

2) Optimisation when the data are imperfectly known

A pollution level is said to be optimal if it is the level for which the marginal costs of treatment and damage are equal, as they are known, estimated or understood by the collective decision-making body. This level is only the optimum for current data and in the framework of the collective decision-making process adopted. Factors which modify the optimal level include any change in the awareness of the public and its representatives, or more comprehensive or accurate data.

Similarly, any change of the decision-making body or of its operating rules could modify the optimal level. As the data required for decisions become increasingly dependent on subjective judgments, it becomes increasingly difficult to know whether a quality criterion adopted by a country corresponds to an optimal or a suboptimal choice. Moreover, a country's choice can generally be taken to be optimal since another choice was not made from the various possibilities.

Any country affected by transfrontier pollution can be a "victim" of wrong judgments on the part of the neighbouring country and in its own interests should provide data to avoid such judgments. Unfortunately, each country can also benefit from altered data. Thus the polluting country could try to reduce the awareness of damage on the part of the victims of pollution, and the polluted country to minimise the implications of adopting a quality criterion for the future economic growth of the other country. In these circumstances, it becomes difficult for one country objectively to claim that the neighbouring country's judgments are wrongful or excessive and to complain about the effects of these judgments on its own economy. Advice from a group of experts nominated by the parties concerned could prove useful but encounters problems if many subjective judgments are involved. However, the countries concerned can restrict the range of possible costs by introducing certain clauses limiting their financial liability in all circumstances, or clauses specifying cancellation and renegotiation in the event of certain limits being exceeded.

Figure 10

IMPLEMENTATION OF AGREEMENT BASED ON MUTUAL COMPENSATION PRINCIPLE
when cost of pollution control can be calculated and cost of damage must be determined
through collective decision-making by the polluted population

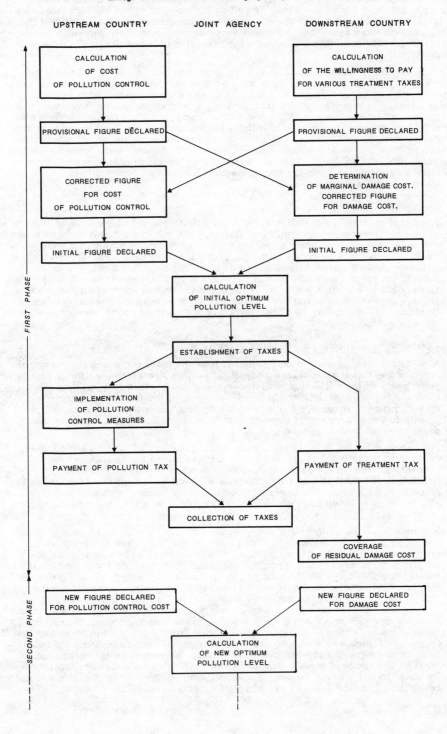

EQUITY

Should the polluting country pay to the joint agency a pollution tax at each pollution level and the polluted country a treatment tax at each treatment level, then each would bear the cost in its own territory and they would pay together the total cost to the joint agency a second time. As such a solution will be seldom accepted by the countries, the joint agency will have to transfer some fixed amount (lump sums) in favour of each country or agree that the countries should pay a tax above a certain level but be granted a bribe below this level. To maintain the utility of the mutual compensation principle, the lump sum transfers (or the levels) will have to be determined such that the amount received by each country (or the level) is unrelated to any inaccuracy which the same country might introduce into the cost estimates it reports (see Annex 1, page 133).

Should the equity problem be dealt with by adopting the Polluter Pays Principle in the strict sense, transfers in favour of the polluting and polluted countries would respectively be the cost of damage at the acceptable pollution level and the cost of pollution control up to this level, as reported to the joint agency and as calculated for the acceptable level under the agreement negotiated between the two countries. It will be seen that in this case the lump sum transfer to each country does not depend on the cost it has estimated.

When the polluter-pays principle and the mutual compensation principle are used together, the polluting country pays a pollution tax to the joint agency when the pollution level exceeds the acceptable level, and receives a bribe if the pollution level is below the acceptable level. The polluted country pays a treatment tax to the joint agency if the pollution level is below the acceptable level and receives a bribe (or compensation) if the pollution level is above the acceptable level. At the acceptable level, neither of the two countries pays or receives anything, but the polluting country does bear the treatment cost up to the acceptable level within its own territory and the polluted country bears the residual damage cost at the acceptable pollution level.

Since the acceptable level is not necessarily the same as the optimal level, and since it is in the interests of both countries to determine the optimal level for pollution, one of the countries will pay the joint agency a tax which will be used to compensate the other country in the light of any change in the pollution level between the acceptable and the optimal levels. In this way, the polluted country will be compensated for the additional damage it suffers if the optimal level is higher than the acceptable level: similarly the polluting country will be compensated for the additional treatment it has to undertake if the optimal level is below the acceptable level. It can be shown that in this particular case the budget of the joint agency will be slightly in deficit and that the transfers from one country to another via the joint agency are nearly in equilibrium only if the acceptable level is close to the optimal level.

The only outstanding question is the way in which the joint agency's probable deficit is financed. It is shown in Annex 1 that the joint agency's budget must not be exactly balanced and that there are methods which make it possible to reduce the unbalance without significant losses of efficiency.

To summarise, the mutual compensation principle can be used together with the polluter-pays principle. In such a case, negotiations between the countries concerned will deal mainly with the acceptable level. If the acceptable level is high, the polluted country will bear high costs and the situation will approach the victim-pays principle. If the acceptable level is low, the polluting country will bear high costs and we approach the civil liability principle.

The mutual compensation principle can also be used in conjunction with any other predetermined principle of equity. For this it is sufficient to determine the lump sum transfers in such a way that each country bears the cost which is wished. Alternatively, it is possible to define the pollution (treatment) level beyond which a country will pay a tax and below which it will receive a bribe. The determination of these levels introduces historical, political, and moral considerations as well as economic and technical factors. The mutual compensation principle does not permit such a choice since it is exclusively an efficiency principle and not at all a principle of equity. For example, there is no transfer to the polluting country, which pays a pollution tax for all emissions if the civil liability principle is adopted. Conversely, there is no transfer to the polluted country, which pays the treatment tax in respect of any treatment carried out by the polluting country if the victim-pays principle is adopted.

If economic and technical situations evolve with time, it is possible to invoke the principle of causality (he who changes pays) (1) by virtue of which any country which deliberately brings about a change in the externality must alone bear the ensuing cost if the other country in no way benefits from the action producing the change. This principle can also be used in conjunction with the mutual compensation principle.

INTERPRETING THE MUTUAL COMPENSATION PRINCIPLE
AS A SCHEME FOR THE SALE OF ENVIRONMENTAL RIGHTS

The mutual compensation principle when the optimal pollution level is reached is equivalent to a system for the sale of certificates or environmental rights between two economic actors (the countries) (see Annex 2). The polluting country buys a limited authorisation to pollute from the joint agency at a price set by the polluted country and paid into the agency. Equilibrium will be reached at the point where the cost of an additional pollution unit equals the cost of an additional unit of treatment undertaken by the polluting country to avoid buying an extra pollution unit. The polluted country buys the right to a specific amount of treatment from the joint agency at a price set by the polluting country and paid into the agency. Equilibrium will be reached when the cost of an additional unit of treatment equals the cost of damage due to an additional unit of pollution which has been accepted by the polluted country in order to avoid the purchase of an extra treatment unit. The polluting country turns over all its pollution rights to the joint agency in exchange for compensation paid by the latter, while the polluted country assigns all its environmental quality rights to the agency in exchange for compensation which the latter also pays.

The joint agency thus acts as a seller of both "pollution rights" to the polluters and of "treatment rights" to the polluted parties, but only acquires this privilege by compensating the countries which originally together held all such environmental rights. In the case of a single country the joint agency and the country are part and parcel of the national community holding all the environmental rights. (2)

An interesting special solution consists in defining a acceptable level of pollution, usually not the optimum level, to give the polluting country the right to pollute up to this level and the polluted country the right to treatment at this same level. Should the polluting country desire to pollute beyond this level, it will have to buy additional rights from the agency at a price set by the polluted country. Should the polluted country require more thorough treatment it will have to buy additional rights from the agency at a price determined by the polluting country. If the acceptable pollution level is above optimum level, the polluted country will

---

1) See report at pages 31-54.

2) In principle, and under certain conditions, a system like that of mutual compensation already exists in a national context when the polluters are subjected to an incentive charge. In this case the polluters explicitly buy limited pollution "rights" (but not the right to pollute) from the community. The victims, who are one with the community, implicitly buy treatment "rights" in that collectively they ultimately bear the costs of treatment, which are reflected throughout the economic system as a collective expenditure for achieving a certain level of pollution regarded as acceptable by the community. In a national context the joint agency is the public authority which collects the charges from the polluters. Through the medium of the budget this authority transfers the amount of tax collected to the victims and sometimes subsidises the polluters. The joint agency is the collective decision-making body of the pollution victims in that the public body is subject to the decisions of the legislative authority, whose powers are conferred by the community which is largely made up of the victims.

Since this system does not sharply distinguish between functions, unsatisfactory situations may sometimes result. Thus if the legislative authority unduly favours the interests of the polluters, the victims may be forced to set up another body which better represents them while the public authority merely plays a mediating role (as in wage negotiations between employers' and workers'

buy treatment rights from the joint agency while the polluting country will sell pollution rights to the agency. This special solution corresponds to the Polluter Pays Principle in the strict sense, evaluated at the acceptable level, but not at the optimal level. It differs from the preceding solution (see page 128) in that the joint agency does not hold the environmental rights, since these remain entirely the property of the two countries, and in that it consequently pays no compensation to the countries.

SUMMARY AND CONCLUSIONS

When costs due to one-way transfrontier pollution are independently estimated by the countries concerned, and the countries have not agreed on such cost estimates, it is shown that there is only one system capable of ensuring economic efficiency. This system is based on the simultaneous taxation of the polluting and polluted parties and gives no indication of how the problem of equity should be solved. Any other system would induce one or both of the parties to provide wrong cost estimates and would hence result in a less efficient solution.

The system proposed has been called the "principle of mutual compensation" to take account of compensations paid or received by the polluters and by the pollutees. It is a symmetrical principle guaranteeing equality of treatment for both countries while favouring neither. Use is made of costs freely evaluated by the countries, while constraints imposed by the notion of national sovereignty are widely met. It is only an economic instrument, since it leaves open the more delicate problem of defining initial rights over the environment or of determining the acceptable level of pollution.

This principle facilitates the solution of existing transfrontier pollution problems as well as those arising from changes in economic situations (creation of polluting or pollution-sensitive activities) or from changes in social preferences when countries manage to agree on a cost-sharing formula meeting some initial situation.

It is particularly well suited to cases where the polluters and the pollutees make up two groups each able to take collective decisions in regard to environmental quality or the acceptable level of pollution. It has the effect of transforming a problem of choice for both groups together into two identical and simultaneous problems of choice for each group individually. Hence the pollution problem posed on an international scale is replaced by two identical problems at national scale (1).

The mutual compensation principle can be used in the event that each country is prepared to provide a monetary estimate of the costs it bears in the first instance (pollution control or damage according to country) and when the countries take rational environmental decisions at national level on the basis of available data.

Cost estimates will be based on all available types of "scientific" data, and recourse for their determination may be had to collective decision-making processes in such a way that national standards, local customs, employment and regional development issues, budget constraints, etc. will be taken into account

---

Note 2) of the page 128 (cont'd):

organisations). Another instance is where the polluters in no way bear the cost of treatment (as when the profits of a polluting enterprise entirely owned by foreigners are only affected). In this event the charges system might result in inefficient decisions, since nothing would prevent the pollutees, who alone would be represented in the collective decision-making body, from demanding an excessive amount of treatment, i.e. from fixing the charge at a high level (although not so high as to force the enterprise to close) because they are not consumers of the enterprise's products and do not implicitly purchase this treatment.

1) A summary of assumptions made in this report is given in Table 1.

Table 1

PRINCIPAL ASSUMPTION OF THE STUDY

A. Pollution model

   (a) One-way (treatment upstream, damage downstream). (Generalisation: Annex 4).

   (b) Static (discounted costs).

   (c) Cost functions to which there corresponds a unique analytical minimum (optimal pollution level).

B. Country model

   (a) Each country resolves individually the questions of income distribution and efficiency.

   (b) Each country works for national efficiency in the light of the available data by minimizing costs.

   (c) Each country can give an evaluation in monetary terms of the costs it bears in the first instance.

   (d) The countries have similar levels of economic development.

C. Model of international relations

   (a) The countries work for global efficiency in the light of the available data.

   (b) The countries have already accepted principles of equity or defined their initial rights.

   (c) The efficient solution and the optimal pollution level do not depend on how the costs are shared between the countries.

   (d) A country cannot transfer the fraction of the costs of transfrontier pollution it bears under an agreement or a de facto situation to another country (no change in prices affecting trade).

   (e) The countries are benevolent (generalisation: see Annex 4).

   (f) No body exists for maximising the collective welfare, nor are there any supranational authorities or redistributive mechanisms between countries.

D. Auxiliary costs

   (a) The costs of information regarding the cost borne in the first instance by a country is charged to this country and considered to be small compared with the gain in efficiency.

   (b) The costs of negotiation and settlement are low compared with the gain in efficiency.

Table 2

ADVANTAGES AND DISADVANTAGES OF THE MUTUAL COMPENSATION
PRINCIPLE

	Advantages	Disadvantages	Remarks
1	Economic efficiency		(a) No other system has this feature in the case in question (b) This is not always a primary objective
2	Does not depend on the equity principle preferred by the countries		Determining the equity principle is a greater problem than that of economic efficiency
3	Reduces an international problem to two national problems		National problems are not always resolved with efficiency in mind
4	(a) Allows sovereign cost estimates (b) Discourages any deliberate error in estimating costs (c) Allows for the social and political aspects of environmental questions	(a) Requires countries to estimate costs they bear in the first instance (b) Introduces the risk of a country falling victim to an error in cost estimate made in good faith by the other country	(a) All efficient solutions require data on costs (b) Maximum amounts borne by each country can be defined
5	No recourse to "experts", courts of international organisations with extensive powers	A joint agency has to be set up	The powers of the joint agency are very limited
6		Taxes are levied, bribes are offered (administrative costs)	It may be reasonable to invoke standards if the administrative costs are too high
7	The pollution level is kept constantly at the optimum	Optimum pollution level not fixed a priori	A priori determination of the optimal level prevents adaptation to new situations
8		The net costs of each country are not determined a priori. The amount of transfers can change with time	The same applies at national level
9		Exchange rate to be defined	Inherent in all international relations

Since the mutual compensation principle is such as to deter any country from reporting costs flawed by any systematic error, countries would be more willing to rely on estimates reached by methods sometimes regarded as "unscientific" whenever data of a more "scientific" kind were lacking (particularly in calculating damage costs). Hence an outstanding advantage of the mutual compensation principle over other systems is that it can easily introduce collective and political cost estimates with no need for these assessments to be reconciled between the two countries.

From a theoretical standpoint the mutual compensation principle appears to be an economic instrument deserving of attention in that it can be implemented provided countries agree on the use of economic instruments as a way of dealing with economic benefit to the countries which adopt it, since information regarding costs borne by the other country will be reduced, while costs resulting from sub-optimisation will be lower; it will also have the political advantage of creating a climate of mutual trust. Moreover, the countries remain entirely free to choose the principle defining their respective initial rights on the environment or the cost sharing formula. Hence it makes no contribution towards solving the problem of equity which dominates negotiations on transfrontier pollution problems.

Implementing the mutual compensation principle requires setting up a joint agency with very limited powers. This agency specifies the optimal pollution level on the basis of information transmitted by the countries but operate under the control of the two countries. It does require neither to call upon an expensive administration nor to delegate extensive powers.

However, the superiority of the mutual compensation principle over other possible instruments (1) does not mean that it should necessarily be invoked in all cases where it might appear justified on theoretical grounds. Indeed, less efficient instruments might be preferable if the advantages of the mutual compensation principle appeared to be outweighed by the difficulties involved in its implementation or acceptance, possibly greater than those of less efficient instruments. However, in all cases the mutual compensation principle does seem to provide a systematic approach likely to enrich other, more traditional methods.

---

1) An outline of advantages and disadvantages of the mutual compensation principle is given in Table 2.

DEFINITION AND ANALYSIS OF THE MUTUAL COMPENSATION PRINCIPLE

INTRODUCTION

This study deals with the special case (1) of one-way transfrontier pollution, where the upstream country pollutes the downstream country and is alone able to treat the effluents. The cost of damage in the downstream country is $D(p)$, where p is the pollution level and the cost of treatment for reducing the pollution to level p is $C(p)$.

The problem of transfrontier pollution is here approached as a "game" (2) between two countries which seek to derive maximum benefit from rules which they will have jointly established while adopting a friendly (co-operative) attitude towards each other. Hence it is first assumed that the countries may sometimes make statements which are not absolutely accurate, and the purpose is to devise a set of "game rules" such as to promote a climate of co-operation where each country has every reason for believing that the other country's statements indeed match the facts. Secondly, it is assumed that the countries show no unfriendly attitude owing to some prevailing climate of hostility and do not try to make the other country bear any higher cost unless this is to their direct advantage (3). This approach is a positive one, even though the initial assumptions may appear excessive.

Game rules in relation to the mutual compensation principle

Each country is assumed to be capable of supplying an inaccurate estimate of the cost on its territory without the knowledge of the other country. Let the inaccurate estimates of treatment costs $C(p)$ and damage costs $D(p)$ be :

$$\alpha C(p) + a \tag{1}$$
$$\delta D(p) + d \tag{2}$$

with α, δ, a and d being independent functions of C, D, and p as well as of three of the four quantities a, d, α and δ. (More complex strategies, where α and δ are functions of p, C or D are not considered, although the same conclusions are reached) (4). In addition it is assumed that quantities α, δ, a and d and functions

---

1) More general cases are dealt with in Annex 4.

2) The "game" approach to transfrontier pollution is described in reports by Messrs. d'Arge, Muraro and Smets, Problems of Transfrontier Pollution, OECD, 1974.

3) The importance of benevolence and malevolence in economic analysis has in particular been discussed by President K.E. Boulding of the American Economic Association ("Economics as a Moral Science", American Economic Review, 59, 1 - 12 (1969).
Reference may also be made to the following article: P. Cazenave and C. Morisson: "Fonctions d'utilité interdépendantes et théorie de la redistribution en économie d'échange", Revue Economique 23, 214-242 (1972).

4) Thus if the upstream country bears $C(p_a) + \delta D(p_a) + d - E_1$, where p_a minimises $[\alpha(p) C(p) + \delta(p) D(p)]$, the net cost for the upstream country is minimal if
$$\frac{\partial}{\partial P_a} \left\{ [1 - \alpha(p_a)] C(p_a) \right\} = 0$$

C(p) and D(p) are such that a single internal solution exists minimising
α C(p) + a + δ D(p) + d (single intersection of marginal cost curves and positive
or non-negative second derivatives).

The "game rules" involving two countries and a joint agency they set up are
as follows:

a) Each country calculates as best it can the cost on its territory for different pollution levels [C(p) or D(p)]. In theory this cost is well known by the country which bears it in the first instance but imperfectly known by the other.

b) Each country reports to the joint agency its estimate of the cost on its territory (α C(p) + a or δ D(p) + d).

c) On the basis of the costs reported the joint agency calculates the pollution level p_a, defined as the level minimising the total reported cost [α C(p) + a + δ D(p) + d]. By hypothesis the level p_a is given by

$$\alpha \frac{\partial C}{\partial P} + \delta \frac{\partial D}{\partial P} = 0$$

and when $\alpha = \delta$ it is always equal to the optimum level p^* minimising the total actual cost T(p) = C(p) + D(p). (1)

d) The upstream country agrees to bear the treatment cost required so that the pollution will not exceed level p_a. The downstream country agrees to bear the cost of damage when the pollution level is equal to level p_a.

e) The upstream country agrees to pay the joint agency a damage cost amount δ D(p_a) + d as reported by the downstream country. The downstream country agrees to pay the joint agency a treatment cost amount α C(p_a) + a as reported by the upstream country. (1)

f) To compensate for these payments the joint agency transfers amount E_1 to the upstream country and amount E_2 to the downstream country, where E_1 does not depend on a, α, p or p_a and where E_2 does not depend on d, δ, p or p_a.

g) If the pollution level is above p_a, the upstream country agrees to pay the downstream country a compensatory amount δ[D(p) - D(p_a)] and the downstream country agrees to make no further claim.

h) If the downstream country requests a pollution level p below p_a, the upstream country agrees to undertake the needed further treatment for a price α [C(p) - C(p_a)] paid by the downstream country to the upstream country (1).

If the pollution level is p_a the net costs borne by the countries are:

Upstream: $\qquad T_M = C(p_a) + \delta D(p_a) + d - E_1(\delta)$ $\qquad\qquad$ (3)

Downstream: $T_V = D(p_a) + \alpha C(p_a) + a - E_2(\alpha)$ $\qquad\qquad$ (4)

The upstream country seeks the best values for a and α so as to minimise the net cost T_M. Since T_M is independent of a, the upstream country has no advantage in increasing the cost of treatment, and since the only effect of such an increase is to add to the cost T_V, it is assumed that the country will choose a = 0 (friendly attitude).

Note 1 of the page 133 (cont'd):

$$\text{i.e., if } [1 - \alpha (p_a)] \frac{\partial C}{\partial p_a} - \frac{\partial \alpha (p_a)}{\partial Pa} C = 0$$

The upstream country minimises its net cost regardless of $p_a(\alpha, \delta)$ when it chooses $\alpha(p_a) = 1$.
The choice of any other solution would appear to be difficult since the upstream does not know the value of p_a (hence of the the functions C(p_a) and $\frac{\partial C(p_a)}{\partial Pa}$) before reporting cost estimate [αC(p) + a].

1) The double taxation system in case of national pollution was proposed by R. H. Coase: "The Problem of Social Cost", Journal of Law and Economics, III, p. 1-44 (1960). See also Reading in Microeconomics (W. Breit and H. M. Hochman, editors), Holt, Rinehart and Winston, New York, 1971.

Figure 1

PENALTIES FOR DECLARING INACCURATE COSTS

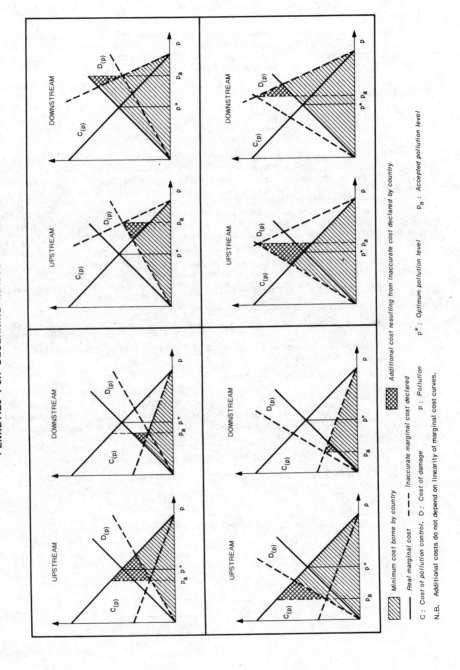

C : Cost of pollution control. D : Cost of damage p : Pollution

N.B. Additional costs do not depend on linearity of marginal cost curves.

Legend:

————— Minimum cost borne by country

————— Real marginal cost

– – – – Inaccurate marginal cost declared

▨ Additional cost resulting from inaccurate cost declared by country

p^* : Optimum pollution level p_a : Accepted pollution level

Since $\dfrac{\partial T_M}{\partial \alpha} = \left[\dfrac{\partial C}{\partial p_a} + \delta \dfrac{\partial D}{\partial p_a}\right] \dfrac{\partial p_a}{\partial \alpha} = \left[\dfrac{\partial C}{\partial p_a} - \alpha \dfrac{\partial C}{\partial p_a}\right] \dfrac{\partial p_a}{\partial \alpha}$,

the upstream country will necessarily choose the value $\alpha = 1$, which minimises its cost T_M (1).

Similarly, the downstream country will choose the value $\delta = 1$ which minimises its cost T_V and the value $d = 0$, since it derives no advantage from an increase in damage cost. A graphical demonstration of the choices $\alpha = \delta = 1$ for increasing marginal costs appears in Figure 1.

The net cost to the upstream country is not minimal with regard to δ for $\alpha = 1$, since:

$$\frac{\partial T_M}{\partial \delta} = (1 - \alpha) \frac{\partial C(p_a)}{\partial p_a} + D(p_a) = D(p_a) > 0$$

The result is a penalty for the upstream country if the downstream country selects a high value for δ (hence for p_a). Owing to the initial postulate of benevolence there is no reason for believing that the downstream country will select a high value for δ since it does not stand to gain by it. Hence the upstream country will pick $\alpha = 1$ whatever the value of δ, while the downstream country will choose $\delta = 1$ regardless of the value of α. The countries will also report the most accurate cost estimates they can and the end result will be that the optimum pollution level is freely chosen by each of the two countries and the joint agency together.

If the pollution level is $p > p_a$ because the upstream country resorts to inadequate treatment, the cost borne by the upstream country will be:

$$T_M(p) = C(p) + \delta D(p) + d - E_1(\delta) \qquad (5)$$

and by the downstream country:

$$T_V(p_a) = D(p_a) + \alpha C(p_a) + a - E_2(\alpha) \qquad (6)$$

The downstream country will select $\delta = 1$ and $d = 0$, hence the upstream country will have to bear a cost

$$T_M(p) = C(p) + D(p) - E_1(\delta)$$

which will always be higher than $T_M(p_a)$.

If this level should be exceeded for any external or accidental reason (assuming that the upstream country treats effluent normally) the upstream country would then bear the net cost

$$C(p_a) + D(p) - E_1(\delta) > T_M(p) > T_M(p_a)$$

It will therefore invariably be induced to take any measure needed for avoiding pollution accidents.

---

1) The second derivative of T_M is positive; thus,

$$\frac{\partial^2 T_M}{\partial \alpha^2} = \frac{\partial^2 C}{\partial p_a^2} (1 - \alpha) \left[\frac{\partial p_a}{\partial \alpha}\right]^2 + \frac{\partial C}{\partial p_a} (1 - \alpha) \frac{\partial^2 p_a}{\partial \alpha^2} - \frac{\partial C}{\partial p_a} \frac{\partial p_a}{\partial \alpha} > 0$$

for $\alpha = 1$ since $\dfrac{\partial C}{\partial p_a} < 0$ and $\dfrac{\partial p_a}{\partial \alpha} = \dfrac{-\dfrac{\partial C}{\partial p_a}}{\alpha \dfrac{\partial^2 C}{\partial p_a^2} + \delta \dfrac{\partial^2 D}{\partial p_a^2}} > 0$.

Since the above "game rules" require the transfers E_1 and E_2 to be defined, the problem at this stage is a matter of calculating the amount of these transfers which will induce the countries to accept an agreement based upon the proposed rules.

If $E_1 = E_2 = 0$, the budget of the joint agency is always beneficiary and the countries each bear the total minimum cost $T(p^*) = C(p^*) + D(p^*)$ where p^* is the optimum pollution level. If the initial situation is such that the upstream country refuses any liability regarding transfrontier pollution, the situation of the downstream country is no better and that of the upstream country apparently worse since it bears the cost $T(p^*)$. It is quite reasonable to expect that the upstream country will refuse this solution as tantamount to adopting a principle of civil liability, and that the downstream country will also turn it down as not providing any appreciable improvement in regard to the victim-pays principle.

Some arguments do however exist in favour of this solution. First of all, if the countries together bear an additional cost $T(p^*)$ (budget of the joint agency), they do not bear any information costs for detecting inaccuracies in the cost estimates by the other country. Moreover, the countries do not bear the costs of administration, supervision and management of the common resource since these charges will be paid out of the budget of the joint agency. The countries might also agree to assign part of the common agency's budget to R and D work on pollution control techniques or methods of reducing damage.

Another possibility would be to assign the accumulated reserves of the joint agency as security against the risk of transfrontier pollution accidents when their cost exceeds the limits covered in the agreement (limited liability of the polluting country).

The unassigned part of the joint agency's budget could possibly be used for projects which, while of no direct benefit to the regions concerned, would in fact be viewed by them in a favourable light. For example such projects might consist in protecting and improving the environment outside these regions, or be voluntary types of action benefiting other parts of the world. This last argument may well not appear very convincing, because it does not seem "normal" for countries affected by transfrontier pollution to give more help to a region of the world, something the countries would not do if they were united.

A better solution would be to allocate lump-sum transfers E_1 and E_2 to the countries as an incentive for accepting the above rules. The downstream country would agree to bear only the cost $C + D - E_2$, a smaller sum than the total cost $C + D$ it would have to bear if the victim-pays principle were applied /or the cost $D(p_{max})$ if no agreement were reached/, and the upstream country should not have to accept any principle of civil liability since it would bear the cost $C + D - E_1$, again less than the total cost $C + D$.

The transfers E_1 and E_2 are both unknowns which must satisfy the two conditions:

$$0 \leqslant E_1 \leqslant T(p^*) \tag{7}$$

$$0 \leqslant E_2 \leqslant T(p^*) \tag{8}$$

and which cannot be paid by the joint agency unless

$$E_1 + E_2 \leqslant T(p^*) \tag{9}$$

Negotiations over the amount of lump sum transfers concern the choice of a point inside the triangle OAB on Figure 2.

The upstream country will prefer point B $/E_1 = T(p^*)/$ and the downstream country, point A $/E_2 = T(p^*)/$. The upstream country will generally be unable to have point B selected, which implies that it need not bear any part of the total cost due to pollution. This attitude would in fact go against the idea of solidarity between the countries and the principle of the liability accepted under international law.

Similarly, the downstream country will generally be unable to have point A selected, since negotiations often begin in a situation which is unfavourable for the downstream country (point B).

However for reasons derived from solidarity it should be possible for the two countries to reach an understanding over the choice of a point located somewhere between A and B. Furthermore, the possibility of international adoption of a principle of complete liability of the polluter sets up a situation of uncertainty for the polluting country, which may well prefer a solution which eliminates risk based upon a principle of partial liability (1).

Figure 2

AREA OF VALUES ALLOWED FOR TRANSFERS E_1 AND E_2

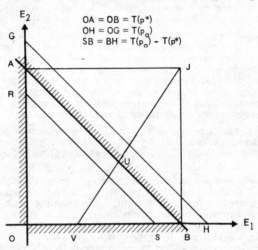

The upstream's choice of the transfer E_1 is linked to the following considerations:

a) The net cost T_M borne by the upstream country must be tolerable and therefore not too high. If $C_S = C(p_S)$ is regarded as the maximum tolerable cost, the transfer E_1 should satisfy the condition

$$E_1 > T(p^*) - C_S \tag{10}$$

b) The upstream country is morally bound to acknowledge that there is a pollution level p_1 beyond which it must accept full liability for the additional damage caused. In other words, the net cost borne by the upstream country when $p \geqslant p_1$ would be at least equal to $C(p) + D(p) - D(p_1) - C(p_1)$. Since the damage cost $D(p_1)$ is usually much higher than the treatment cost $C(p_1)$, the transfer E_1 should satisfy the condition $E_1 \leqslant \delta D(p_1) < \delta D(p_1) + \alpha C(p_1)$. The pollution level p_1 could be a maximum pollution level adopted internationally having regard to "normal" practices in the region or an "average" pollution level for the countries concerned.

The downstream country's choice of the transfer E_2 is linked to the following considerations:

a) The net cost T_V borne by the downstream country must be tolerable and therefore not too high compared with the income of the region in question. If D_t is the tolerable damage, the transfer E_2 should satisfy the condition:

$$E_2 > T(p^*) - D_t \tag{11}$$

1) See papers by H. Smets and G. Muraro in Problems of transfrontier pollution, OECD, 1974.

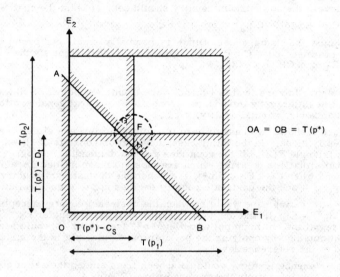

Figure 3a

SITUATION CORRESPONDING
TO LARGELY INCOMPATIBLE STANDPOINTS

OA = OB = T (p*)

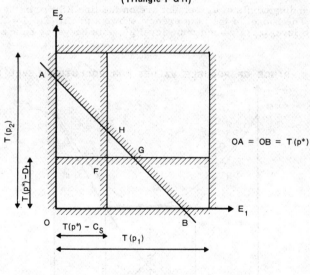

Figure 3b

SITUATION CORRESPONDING TO COMPATIBLE STANDPOINTS
(Triangle F G H)

OA = OB = T (p*)

In fact, the tolerable damage will be linked to the damage suffered by the inhabitants of the other regions of the country and therefore to the maximum pollution level regarded as nationally or internationally tolerable (p_t)

$$D_t \approx D(p_t)$$

b) The downstream country is usually bound to acknowledge that there is a "normal" quality level and that it cannot demand additional treatment without bearing the extra cost involved. If p_2 is the "normal" pollution level, the downstream country should bear at least the cost

$$C(p) + D(p) - C(p_2) - D(p_2) \text{ if } p < p_2.$$

Since the damage cost $D(p_2)$ is normally small compared to the treatment cost $C(p_2)$, the transfer E_2 should satisfy the condition $E_2 \leqslant \alpha C(p_2) < \alpha C(p_2 + \delta D(p_2)$.

The above factors tend to reconcile the countries' points of view and reduce the area of negotiation by introducing economic or general considerations (international standards, normal quality, etc.).

Figure 3 shows two cases where the points of view are either very divergent or sufficiently close. In the latter case, negotiation consists of choosing a point inside the triangle FGH. If the countries are too demanding, no agreement is possible but situations do arise which are close to those needed for agreement. This happens when $C_S + D_t < T(p^*)$, i.e. when the downstream country chooses the level $p_t \approx 0$ and the upstream country decides not to reduce pollution.

One interesting case would be when the upstream country accepts the transfer $E_1 = T(p^*) - C(p_0)$, and the downstream country the transfer $E_2 = T(p^*) - D(p_0)$. In this case, at the optimum pollution level p^*, the countries would bear exactly the amount obtained by applying the polluter-pays principle in the strict sense calculated at the reference level p_0.

The reference level p_0 would be chosen from among the set of pollution level values p which satisfy the two conditions (Figure 4).

$$C(p) < T(p^*) \tag{12}$$

$$D(p) < T(p^*) \tag{13}$$

The corresponding points of agreement are on the line RS within the triangle OAB (Figure 2). The budget of the joint agency shows a profit $\underline{/}T(p_0) - T(p^*) > 0\underline{/}$. If the reference level p_0 is the optimum level p^*, this budget is balanced.

Figure 4

RANGE OF POSSIBLE VALUES FOR POLLUTION LEVEL Po

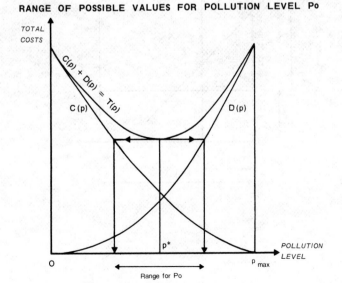

140

A similar case is that in which the transfers are

$$E_1 = T(p_o) - C(p_o) = D(p_o) \text{ and } E_2 = T(p_o) - D(p_o) = C(p_o).$$

These values for E_1 and E_2 correspond to a point on line GH of Figure 2.
The countries then bear the net costs

$$T_M = C(p_o) - \left[T(p_o) - T(p^*) \right] < C(p_o)$$

$$T_V = D(p_o) - \left[T(p_o) - T(p^*) \right] < D(p_o)$$

when the pollution level is the optimum level p^*, and the net costs

$$T_M = C(p_o)$$

$$T_V = D(p_o)$$

when the pollution level is the reference level p_o (polluter pays principle in the strict sense calculated at the reference level p_o). In this case, the budget of the joint agency

$$T(p^*) - \left[E_1 + E_2 \right] = T(p^*) - T(p_o) < 0$$

shows a deficit which grows with the difference between the reference level and the optimal level.

This case is nonetheless interesting, since it reflects a situation in which the countries define a reference (or acceptable) level p_o after agreeing on the principle that transfers are given by

$$E_1 = \delta D(p_o) \tag{14}$$

$$E_2 = \alpha C(p_o) \tag{15}$$

In such cases, countries must only provide cost differentials $\delta \underline{/} D(p) - D(p_o) \underline{/}$ and $\underline{/} C(p) - C(p_o) \underline{/}$.

Negotiations might well concentrate more on the idea of equity than on the reference level, the economic implications of which are not always easily understood by non-specialists. Generally speaking, the net costs T_M and T_V are considered to be "fairly" distributed when

$$\frac{T_M}{B_M} = \frac{T_V}{B_V} = \frac{T_M + T_V}{B_M + B_V} \tag{16}$$

where B_M and B_V are characteristic values for each region or country (for example, income or population of the regions or countries). If $x = B_M/B_V$, the transfers E_1 and E_2 must satisfy the condition:

$$T(p^*) - E_1 = x \left[T(p^*) - E_2 \right] \tag{17}$$

The relevant values of E_1 and E_2 lie on a straight line UV and the countries will choose the point U if they are able (Figure 2). Thus the adoption by the countries of a principle of equity removes any need for negotiation regarding the transfers E_1 and E_2, when the total cost $T(p^*)$ is known.

In reality, choice of the transfers E_1 and E_2 is complicated because it must be made before the costs of treatment (C) and damage (D) and the total cost (T) are declared and because each country can have but an approximate idea of the real cost incurred first in the other country. The countries will therefore be induced to choose the values E_1 and E_2, which overall may prove to be too high (the budget of the joint agency showing a deficit) or even unfair (if one of the countries believes that the cost in the other country is very different from that declared). The problem of negotiation thus really takes on the nature of a "game", the aim of which is to acquire the greatest possible share of the joint agency's budget.

For instance, if the countries negotiate on the basis of formulas (14) and (15), and if the reference level p_o is considered as a fixed quantity quite outside the negotiations, there will be no incentive for the countries to have the other country overestimate the cost they declare since the transfer received will not depend on the cost declared. Consequently the transfers E_1 and E_2 should reflect estimates which are as accurate as possible.

Conversely, if the countries believe that the reference level p_o should be close to the optimum level, there is an incentive for the upstream (downstream) country to suggest that the cost of treatment (damage) to be declared is high in order to profit from a transfer $\bar{E}_1(\bar{E}_2)$ which is as high as possible. The country to benefit will be the one which succeeds in having the largest overestimate accepted.

If the reference level p_o is chosen so that the transfers $\bar{E}_1$ and $\bar{E}_2$ are approximately equal when the countries will declare the costs, the upstream country will try to suggest that the cost of treatment to be declared is high in order to obtain a high level p_o and therefore a high transfer $\bar{E}_1$. The downstream country will try to suggest that the cost of damage to be declared is high in order to obtain a low reference level p_o and therefore a high transfer $\bar{E}_2$.

Situation in which the joint agency's budget shows a deficit

If the budget of the joint agency shows a slight surplus, there will be little problem in disposing of this surplus. On the other hand, if the budget shows a deficit, some special mechanism will have to be agreed upon. This will happen if the countries believe the total cost $T(p^*)$ to be higher than it actually is, and if they choose transfers $\bar{E}_1$ and $\bar{E}_2$ such that $\bar{E}_1 + \bar{E}_2 > T(p^*)$.

In this case the countries could choose new values $\bar{E}_1^1$ and $\bar{E}_2^1$ such that the budget is balanced, i.e.

$$\bar{E}_1^1 + \bar{E}_2^1 = \alpha\, C(p_a) + \delta\, D(p_a) \tag{18}$$

where $p_a = p_a(\alpha, \delta)$

Such a solution would be satisfactory if it were possible to choose transfers $\bar{E}_1^1$ independent of α and $\bar{E}_2^1$ independent of δ. Such a choice generally appears impossible because the sum $\alpha\, C(p_a) + \delta\, D(p_a)$ cannot be written in the form $f(\alpha) + g(\delta)$. Thus it is no longer possible to reach an exact estimation of costs once the budget has to be balanced.

Example: if $E_1^1 = \delta\, D(p_a)$ and $E_2^1 = \alpha C(p_a)$ /classical polluter-pays principle, see Annex 2/, we have

$$T_M = C(p) + \delta D(p) - \delta\, D(p_a)$$

$$T_V = D(p) + \alpha C(p) - \alpha C(p_a)$$

The upstream country chooses the value of α which minimises $T_M(p_a)$.

$$\text{Since } \frac{\delta\, T_M}{\delta\,\alpha} = \frac{\delta C(p_a)}{\delta\, p_a}\, \frac{\delta p_a}{\delta\,\alpha} \tag{19}$$

$$\frac{\delta\, C(p_a)}{\delta\, p_a} < 0 \text{ and } \frac{\delta p_a}{\delta\,\alpha} > 0,$$

T_M decreases as α increases. Similarly T_V decreases as δ increases.

It is in the interest of both countries to exaggerate the costs.

Special case: Transfers when the countries have agreed upon an acceptable level of pollution

If the countries have agreed upon the polluter-pays principle at the level p_o, the net costs they bear when they apply the mutual compensation principle should be such that

$$T_M = C(p^*) + D(p^*) + D(p_o) + E_1'' \leqslant C(p_o)$$

$$T_V = D(p^*) + C(p^*) + C(p_o) + E_2'' \leqslant D(p_o)$$

142

or, in other words, $E_1'' \leqslant T(p_0) - T(p^*)$

$$E_2'' \leqslant T(p_0) - T(p^*)$$

The budget of the joint agency will not be in deficit if:

$$E_1'' + E_2'' \geqslant T(p_0) - T(p^*)$$

These three conditions define the zone of possible values for the transfers E_1'' and E_2''. If the transfers E_1'' and E_2'' are to be equal, they must be between $\frac{1}{2} \underline{/}T(p_0) - T(p^*)\underline{/}$ and $T(p_0) - T(p^*)$. If the countries choose E_1'' and E_2'' before the costs are announced, they may take too high a value if they overestimate the surplus $T(p_0) - T(p^*)$ or too low a value if they underestimate this surplus.

If the budget has to be balanced, one solution is to ask "other" countries to balance the budget, i.e. to pay the sum

$$\overline{E}_1 + \overline{E}_2 - T(p^*)$$

In this case, there is no point in the countries concerned by transfrontier pollution supplying an inaccurate estimate of costs. This solution is not realistic because the "other" countries have no reason to give an advantage to the countries involved.

A better solution is to set up a joint fund between n pairs of countries concerned by n problems of transfrontier pollution. This fund, which could be financed by a tax, would balance the budgets of the n joint agencies. In this case, the upstream and downstream countries would bear the cost.

$$T_M = C(p) + \delta D(p) - \overline{E}_1 + \left\{ \beta_1 \left[\overline{E}_1 + \overline{E}_2 - \alpha C(p) - \delta D(p) \right] + B_1 \right\} \qquad (20)$$

$$T_V = D(p) + \alpha C(p) - \overline{E}_2 + \left\{ \beta_2 \left[\overline{E}_1 + \overline{E}_2 - \alpha C(p) - \delta D(p) \right] + B_2 \right\} \qquad (21)$$

where $\beta_1 \approx 0$, $\beta_2 \approx 0$, $B_1 > 0$, $B_2 > 0$

The fractions β_1 and β_2 are inversely proportional to n. The amounts B_1 and B_2 are related to the payments made from the joint fund to the (n - 1) other joint agencies. The terms in curly brackets in the equations (20) and (21) represent the taxes paid into the joint fund. This solution becomes even better (1) as the number n increases. It is only acceptable to the upstream and downstream countries if the tax paid into the joint fund is not too high. For example, if $\overline{E}_1 = \delta D(p_0)$ and $\overline{E}_2 \approx \alpha C(p_0)$, the optimal solution releases a surplus $T(p_0) - T(p^*)$ for the two countries. The taxes paid into the fund by the two countries must be less than this surplus. This kind of condition can always be met if there are n similar cases of transfrontier pollution with

$$\beta_1 = \beta_2 = \frac{1}{n} \quad \text{and}$$

$$\beta_1 = \beta_2 = \frac{n-1}{n} \quad \underline{/}T(p_0) - T(p^*)\underline{/}$$

It could not be met if the cases are dissimilar because the quantities β_1, β_2, B_1 and B_2 do not reflect this difference which is difficult to calculate exactly.

Another solution would be to arrange for the deficit to be borne by future generations in the two countries concerned.

For this to be possible, the country taking the initiative simply makes little allowance for repaying the deficit it is creating (i.e. a short-sighted approach or a short-term decision). In other words, the discount rate appearing in the calculation done by the decision-maker would have to be appreciably larger than the rate of interest. This solution is inapplicable in practice because countries are unaware of the different rates used by others.

1) The influence of this tax on decision-making can be zero when the decision regarding transfrontier pollution is taken by a ministry and the international tax, under the control of another ministry, is lumped together with other international taxes (payments to international organisations).

However if the countries delegate the decision regarding costs to the regions affected by the transfrontier pollution, and undertake to make interest-free loans to the joint agency, the result is the same. Ultimately, the region affected by transfrontier pollution is not liable for the joint agency's deficit and acts as if this were zero, since the deficit is divided between all the regions of the country.

The region which benefits - if it is small with respect to the country - bears a small proportion of the deficit covered by the national budget and pays additional taxes when its wealth increases as a result of optimisation.

If the rate of the additional taxes approaches 50 per cent, the national budget will not be affected by the mechanism described (see Table below). The region which benefits will finally increase its well-being and will supply the best estimate of the cost (insofar as it takes no account of the taxes in its calculation).

	Variation in costs				
	Polluting country excepting the region where the pollution originates	Region where the pollution originates	Polluted region	Polluted country excepting the polluted region	Joint agency
Initial level: p_O	0	0	0	0	0
Optimal level: p^*					
a) mutual compensation	0	$+ S$	$+ S$	0	$- S$
b) interest-free loan	$- \dfrac{S}{2}$	0	0	$- \dfrac{S}{2}$	$+ S$
c) repayment of the loan with the interest credit shared between regions	$+ a\dfrac{S}{2}$	$- a\dfrac{S}{2}$	$- a\dfrac{S}{2}$	$+ a\dfrac{S}{2}$	
d) taxes on the surplus	$+ bS$	$- bS$	$- bS$	$+ bS$	
Total	$(\dfrac{a}{2}+b-\dfrac{1}{2}) S$	$(1-\dfrac{a}{2}- b) S$	$(1-\dfrac{a}{2} - b) S$	$(\dfrac{a}{2}+b - \dfrac{1}{2}) S$	0

If $a \approx 0$, $b \approx 50$ per cent, the remainder of the country does not give a net advantage to the region concerned in the transfrontier pollution.

If $(\dfrac{a}{2} + b) < \dfrac{1}{2}$, the international agreement also involves the internal transfer of aid to the region involved in the transfrontier pollution.

With the solution that consists in financing the joint agency by a fixed tax defined a priori or a capital grant, the problem of the deficit can be resolved in the short term. In the long term however this solution has the disadvantage that the countries are apprehensive about having to make a further capital grant or having to change the tax if the initial grant or tax should prove insufficient.

To sum up, there is no perfect way of balancing the budget without sacrificing efficiency. If stress is laid on balancing the budget (or on equity), efficiency cannot be achieved in a rigorous manner. However if a small loss of efficiency is acceptable, then practical systems which balance the budget are conceivable.

Incidentally, the budget of the joint agency may go rapidly into surplus when the transfers $\overline{E}_1$ and $\overline{E}_2$ are fixed but the total cost $T = C + D$ increases with time (see page 145). This situation can arise when the damage cost increases (greater awareness and larger exposed population) and when the treatment cost increases (the effect of greater industrialisation not compensated by improved pollution control techniques).

UNIQUENESS OF THE PROPOSED SYSTEM

The proposed system is unique if there are no additional lump sum transfer E_3, E_4, which modifies the above rules and leads to exact estimates ($\alpha = \delta = 1$) of the costs declared by the countries.

The special situation where the transfers E_3 and E_4 are linked to the declared costs is here considered: $E_3 = E_3(\alpha C, \delta D)$, $E_4(\alpha C, \delta D)$. In order that the upstream country choose the value $\alpha = 1$, E_3 should be such that:

$$\frac{\partial E_3}{\partial(\alpha C)}\left[\alpha \frac{\partial C}{\partial p}\cdot\frac{\partial p}{\partial \alpha}+C\right]+\frac{\partial E_3}{\partial(\delta D)}\,\delta\,\frac{\partial D}{\partial p}\,\frac{\partial p}{\partial \alpha} = 0$$

or

$$\frac{\dfrac{\partial E_3}{\partial(\delta D)}-\dfrac{\partial E_3}{\partial(\alpha C)}}{\dfrac{\partial E_3}{\partial(\alpha C)}} = \frac{C}{\alpha\dfrac{\partial C}{\partial p}\dfrac{\partial p}{\partial \alpha}} = \frac{-\alpha C\left[\alpha\,\dfrac{\partial^2 C}{\partial p^2}+\dfrac{\partial^2 D}{\partial p^2}\right]}{\left[\alpha\,\dfrac{\partial C}{\partial p}\right]^2} \qquad (22)$$

Equation (22) cannot be satisfied irrespective of the values of C and D because E_3 is not a function of

$$\alpha\,\frac{\partial C}{\partial p}\,,\quad \alpha\,\frac{\partial^2 C}{\partial p^2}\quad \text{and}\quad \delta\,\frac{\partial^2 D}{\partial p^2}$$

Thus the hypothesis that there exists a non-zero value of E_3 is absurd. Similarly, it may be shown that E_4 equals 0. In Annex 2 it is shown directly that other existing systems (E_3 or $E_4 \neq 0$) do not lead to efficiency.

Hence the mutual compensation principle would be the only method which can both result in economic efficiency and accurate cost estimates. This proof could be generalised to cases where E_3 and E_4 are functions of αC, δD, $\alpha\dfrac{\partial C}{\partial p}$, $\delta\dfrac{\partial D}{\partial p}$, etc.

VARIATION OVER TIME

If the costs change over time are:

$$C = \left[1+x(t)\right]C(p)$$
$$D = \left[1+y(t)\right]D(p)$$

where $x(0) = y(0) = 0$, then the optimum pollution level will also change over time. If the transfers E_1 and E_2 are constant the net costs in the two countries and the joint agency will equally vary.

If $E_1 = \left[1+y(t)\right]D(p_1)$ and $E_2 = \left[1+x(t)\right]C(p_2)$, the cost to the upstream country will undergo a smaller change when the damage cost varies, and the downstream country a smaller change when the cost of treatment varies. Thus:

$$\frac{\partial T_M}{\partial y} = D\left[p_a(x, y)\right] - D(p_1) < D(p_a)$$

$$\frac{\partial T_V}{\partial x} = C\left[p_a(x, y)\right] - C(p_2) < C(p_a)$$

145

For the net cost borne by a country to be independent of time changes in the other country's cost, the quantities p_1 and p_2 would have to vary with p_a, which seems difficult to achieve. The levels p_1 and p_2 could however be chosen so that the derivatives $\dfrac{\partial T_M}{\partial y}$ and $\dfrac{\partial T_V}{\partial x}$ average out to zero so long as the functions $x(t)$ and $y(t)$ could be predicted.

A radically different solution would be to accept the rule that any country which changes the costs on its territory by virtue of economic decisions must bear all the additional net cost, the other country's net costs remaining unchanged (causality principle) (1). If the upstream country develops a polluting industry such that the optimum pollution level changes from $p_o{}^*$ to $p_t{}^*$ it would have to transfer to the downstream country the additional total cost $T_t(p_t{}^*) - T_o(p_o{}^*)$ generated by establishing a new polluting industry. The joint agency would thus maintain a constant budget. This solution would have the advantage of preventing the creation of polluting industry in any suitable region, since the upstream country would not set it up unless the resulting increase in income was greater than the increase in total cost $T_t(p_t{}^*) - T_o(p_o{}^*)$. The same reasoning applies to the case where the downstream country initiates some pollution-sensitive activity.

If the cost of pollution control measures $C_o(p)$ becomes $C_t(p)$ (new pollution industry), the cost of damage remains unchanged and equal to $D_o(p)$ a function which was declared as accurately as possible during earlier periods. The downstream country cannot declare a different damage cost since there has been no change in its situation. It does not bear a change in net cost because the upstream country provides him with an indemnity equal to the change in net cost. Then the upstream country is subject to a pollution tax and the downstream country is indifferent to the selection of the level p^*t. The upstream country will declare the cost $\alpha C_t(p)$ and will choose the value $\alpha = 1$ since the cost of pollution control is the only one not well-known and the "taxed" country is the one which knows it best (application of the cost-sharing principles discussed in the report at pages 87-114).

To summarise the following approach would seem worth considering in order to take account of changes over time and to preserve the validity overtime of any agreement based on the mutual compensation principle:

a) choose values of the transfers E_1 and E_2 which account for time variations over time rather than fixed amounts: (e. g. the polluter pays principle at the reference level p_o) ;

b) apply the causality principle ("he who changes must pay") only for significant changes resulting from unarguably economic decisions;

c) include a clause to the effect that the agreement will be renegotiated if very large and unexpected changes should take place (for example, very cheap technical methods of pollution control);

d) arrange for the agreement to apply at least over some definite but sufficiently long period (for example 25 years).

---

1) The causality principle (see report at pages 31-54) does not apply when the change in costs is due to external factors such as improved knowledge about damage or the development of better methods of pollution control. It only applies in cases where countries benefit from increased income as a result of changes made to the economic situation (investment, land-use schemes, geographical distribution of industry, etc.). In practice, it will be necessary to distinguish from among the cost changes declared by the countries those coming under the causality principle. The countries will have to agree on this point, since under the causality principle the country making a change must bear a higher cost. This agreement will probably cover large projects (irrigation, changing the course of rivers, building new towns, establishing industrial zones or port facilities, creating nature reserves, etc.) and will take no account of small and gradual changes in polluting or pollution-sensitive activities (for example small increases in population, changes in consumer tastes and habits).

Additional Note:

The Mutual Compensation principle which was submitted to the OECD Sub-Committee of Economic Experts in April 1973 gave rise to further studies within the context of "game theory" by:

Cl. d'Apremont and L.A. Gérard-Varet (CORE, Louvain):

- "Un modèle de solution négociée des problèmes de pollution transfrontière unidirectionnelle", report CB/1/11 in "Modèles de décision pour le choix de la qualité des eaux et la répartition des charges de l'épuration", December 1974, Working Group on Environmental Economics, Université Catholique de Louvain (to be published in Revue d'économie politique).

- "Cost sharing and incentives in a pollution game with incomplete information", report CB/1/12 in the same document, to be published in an international periodical.

- "Individual incentives and collective efficiency for an externality game with incomplete information", CORE discussion paper 7519 (July 1975).

Annex 2

BRIEF EXAMINATION OF OTHER INSTRUMENTS

The purpose of this annex is to give a brief description of various other economic instruments (1) which may be used in connection with transfrontier pollution and to demonstrate that these instruments are not economically efficient. This description supplements that given in Annex 1 where it was shown that the mutual compensation principle was the only instrument having the required characteristics.

This Annex is in three sections. The first section examines two-**party** instruments while Section 2 looks at those involving in addition a joint agency with an unbalanced budget. Finally, results are given from a comparative calculation of solutions for an imaginary situation. In each case, it is assumed that declared costs differ from true costs because of a multiplicative constant. There are of course other strategies especially when countries can make rough estimates of the costs borne by the other country.

A. TWO-PARTY SYSTEMS

If the polluter initially has the right to discharge pollutants until the pollution level p_o is reached, and if the polluted party is able to sell the polluter permission to pollute $(p > p_o)$ or to buy from the polluter the service of reduced pollution $(p < p_o)$, the polluter and pollutee together own all the environmental rights and together bear the cost of treatment and residual damage. Negotiations between the two parties lead to the optimum solution if the total cost due to pollution (treatment and damage) is minimal. The sum of the costs borne by the two parties is equal to the total cost.

The conventional polluter-pays principle

If game rules are adopted according to which the polluting country bears the cost of pollution control in order to reduce pollution to an acceptable level and the polluted country the damage cost up to this acceptable level, the costs borne by the countries are:

$$\begin{cases} T_M = C(p_a) \\ T_V = D(p_a) \end{cases}$$

If the countries declare costs $\alpha C(p)$ and $\delta D(p)$, the "acceptable" pollution level p_a is chosen so as to minimise the total declared cost. In this case, p_a is a solution of:

$$\alpha \frac{\partial C}{\partial p} + \delta \frac{\partial D}{\partial p} = 0$$

The polluting country which chooses the maximum value of α since T_M is a decreasing function of p_a which itself is an increasing function of α. Similarly, the polluted country will choose the maximum value of δ. This game rule encourages the most flagrant exaggerations in costs and benefits the country which manages to get the highest over-estimate accepted. This rule only produces the optimum solution if the countries are able to give equally inaccurate estimates and results in a climate being set up which is hardly propitious to a harmonious solution of problems of transfrontier pollution.

---

1) A graphical presentation of various economic instruments is given in the report at pages 55-86.

The modified victim-pays principle (MVP) and the modified civil liability principle (MCL)

As indicated in report at pages 87-114, the net costs borne by the countries according to the modified victim-pays principle and the modified civil liability principle respectively are:

$$\text{MVP} \quad \begin{cases} T_M = C(p_o) + (1-\alpha)\left[C(p) - C(p_o)\right] \\ T_V = D(p) + \alpha\left[C(p) - C(p_o)\right] \end{cases}$$

and

$$\text{MCL} \quad \begin{cases} T_M = C(p) + \delta\left[D(p) - D(p_o)\right] \\ T_V = D(p_o) + (1-\delta)\left[D(p) - D(p_o)\right] \end{cases}$$

where p_o is the reference level. The optimum level p^* is the level which minimises the total cost $T = C+D$.

In the first case (MVP), it is in the interest of the polluting country to declare inaccurate treatment cost because the cost that country bears (T_M) is a minimum for

$$\alpha = 1 + \frac{C(p_a) - C(p_o)}{\dfrac{\partial p_a}{\partial \alpha} \left| \dfrac{\partial C(p_a)}{\partial p} \right|}$$

where p_a is the value of pollution level which minimises the cost borne by the polluted country (T_V). Thus if the countries have not agreed about the treatment cost, this system is unsuitable and encourages exaggeration ($\alpha > 1$) if $p_a < p_o$. By choosing the reference level p_o close to the acceptable level p_a, it is possible to reduce the divergence from the optimum solution. However, this system is ideal when the damage cost is the only cost which is difficult to estimate.

In the second case (MCL), it is in the interest of the polluted country to declare a higher damage cost than is in fact the case if $p_o < p_a$ since the cost the polluted country bears (T_V) is a minimum for

$$\delta = 1 + \frac{D(p_a) - D(p_o)}{\dfrac{\partial D(p_a)}{\partial p} \left| \dfrac{\partial p_a}{\partial \delta} \right|} > 1$$

where p_a is the pollution level which minimises the cost borne by the polluting country T_M. If the countries have not agreed on the damage cost, this system is unsuitable, however it is ideal when the treatment cost is the only cost which is difficult to estimate.

Sharing treatment costs (1)

If the polluting country has to bear a fixed fraction f of the treatment cost and the polluted country the remainder, if the polluting country has to keep the pollution level below p_1 and declares a treatment cost α $C(p)$ to the polluted country, then the costs borne by the polluting country and the polluted country at the level p_1 are

$$T_M = C(p_1) - (1-f)\,\alpha C(p_1) = fC(p_1) + (1-f)(1-\alpha)C(p_1)$$

$$T_V = (1-f)\,\alpha C(p_1) + D(p_1)$$

---

1) See papers by Muraro and Smets in "Problems of Transfrontier Pollution", OECD, 1974.

Since p_1 is usually higher than the pollution level which minimises the cost borne by the polluted country (T_V), the polluted country will buy the service of additional treatment for a price $\alpha C(p) - \alpha C(p_1)$ and will choose the level p_a which minimises the cost $\alpha C(p) + D(p)$. On this basis, the net cost of the polluted country is $T_V = \alpha C(p_a) - f \alpha C(p_1) + D(p_a)$ and the net cost of the polluting country:

$$T_M = (1 - \alpha) C(p_a) + f \alpha C(p_1) = fC(p_1) + (1 - \alpha) \left[C(p_a) - fC(p_1) \right]$$

is a minimum with respect to α if

$$\alpha - 1 = \frac{C(p_a) - f C(p_1)}{\left| \dfrac{\partial C(p_a)}{\partial p} \quad \right| \quad \dfrac{\partial p_a}{\partial \alpha}} > 0$$

Consequently, it is in the interest of the polluting country to exaggerate ($\alpha > 1$) the treatment cost. This system is similar to that based on the modified victim-pays principle in the case in question $\left[fC(p_1) = C(p_0) \right]$ and has the same advantages and disadvantages.

The certificate system

The system based upon the creation of certificates entitling the holder to discharge a certain amount of pollutants was suggested by Dales and examined by a number of authors. Scott in particular suggested the use of this system for problems of transfrontier pollution (1). The following section examines the advantages of this system in different conditions of competition between the polluters and the polluted countries.

If the polluting country is assumed to hold the certificates it needs for a certain pollutant emission corresponding to the pollution level p_0 and the polluted country to hold all the other available certificates (bilateral monopoly situation), the polluting country bears the net cost $T_M = C(p) + \delta \left[D(p) - D(p_0) \right]$ if it buys $(p > p_0)$ additional certificates from the polluting country at the price $\delta \left[D(p) - D(p_0) \right]$ and the net cost

$$T_M = C(p) + \alpha \left[C(p_0) - C(p) \right] = C(p_0) + (1 - \alpha) \left[C(p) - C(p_0) \right]$$

if it sells $(p < p_0)$ certificates at the price $\alpha \left[C(p) - C(p_0) \right].$

The polluted country bears the net cost:

or

$$T_V = D(p) + \delta \left[D(p_0) - D(p) \right] = D(p_0) + (1 - \delta) \left[D(p) - D(p_0) \right] \text{ if } p > p_0$$

$$T_V = D(p) + \alpha \left[C(p) - C(p_0) \right] \text{ if } p < p_0.$$

The certificate system produces the same result as the modified victim-pays principle $(p < p_0)$ or the modified civil liability principle $(p > p_0)$.

If $p > p_0$, the polluting country buys authorisation from the polluted country to pollute more. The acceptable level of pollution minimises the net cost of the polluting country but the polluted country will declare a damage cost greater than reality since that country's net cost is a minimum for

$$\delta - 1 = \frac{D(p_a) - D(p_0)}{\dfrac{\partial D(p_a)}{\partial p} \left| \dfrac{\partial p_a}{\partial \delta} \right|}$$

where p_a is a solution of

$$\delta \frac{\partial D(p)}{\partial p} + \frac{\partial C(p)}{\partial p} = 0$$

---

1) See Annex 2 to the paper by Scott and Bramsen in "Problems of transfrontier pollution", OECD, 1974.

Finally the polluting country will be induced to carry out treatment to a level above the optimum $(p_a < p^*)$.

If $p < p_0$, it can be shown similarly that the polluted country will choose too low a level of treatment $(p_a > p^*)$ because it will be in the interests of the polluting country to declare a higher treatment cost

$$\alpha - 1 = \frac{C(p_a) - C(p_0)}{\left|\dfrac{\partial C(p_a)}{\partial p}\right| \dfrac{\partial p_a}{\partial \alpha}} > 0$$

where p_a is a solution of

$$\alpha \frac{\partial C(p)}{\partial p} + \frac{\partial D(p)}{\partial p} = 0$$

To conclude, the certificate system always encourages declaration of costs different from reality and does not lead to the most efficient solution.

Remarks

a) The certificate system is similar to that based upon the mutual compensation principle when the reference pollution level p_0 is close to the optimum level p^*. For if

$$C(p^*) - C(p_0) \approx D(p_0) - D(p^*), \text{ the net costs are:}$$

$$T_M \approx C(p^*) + \delta \left[D(p^*) - D(p_0)\right] \approx C(p^*) + \alpha \left[C(p_0) - C(p^*)\right]$$

$$T_V \approx D(p^*) + \alpha \left[C(p^*) - C(p_0)\right] \approx D(p^*) + \delta \left[D(p_0) - D(p^*)\right]$$

b) The certificate system could be converted into a system equivalent to that based upon the mutual compensation principle if an equalisation fund were introduced to increase the price paid for the certificates.

If $p > p_0$, the polluting country would pay $\delta\left[D(p) - D(p_0)\right] > 0$ and the equalisation fund $\alpha\left[C(p_0) - C(p)\right] + \delta\left[D(p_0) - D(p)\right] > 0$ so that the polluted country receives $\alpha\left[C(p_0) - C(p)\right]$ i.e. a higher price for certificates sold. If $p < p_0$, the polluted country would pay $\alpha\left[C(p) - C(p_0)\right] > 0$ and the equalisation fund $\alpha\left[C(p_0) - C(p)\right] + \delta\left[D(p_0) - D(p)\right] > 0$ so that the polluting country receives $\delta\left[D(p_0) - D(p)\right] > 0$ i.e. a higher price for the certificates sold. When the reference level p_0 is close to the optimum level $(p^* \approx p_0)$, the equalisation fund serves no further purpose and the certificate system is comparable to that based on the mutual compensation principle.

c) If there is only a single polluter who has to buy certificates $(p > p_0)$ from the polluted parties in a situation of perfect competition, the polluted parties (1) will receive $D(p) - D(p_0)$ and the polluters will obtain the certificates at their true cost. If the polluters sell certificates $(p < p_0)$, the polluted parties will have to pay $D(p_0) - D(p)$, which is higher than $C(p) - C(p_0)$, because of competition. In this case, the effect of competition is to transfer the optimisation gain to the polluter alone, to cause the polluted parties to declare the exact damage cost and to keep the net cost of the polluted parties at a constant value. The certificate system corresponds to the principle of taxing polluters (see report at pages 87-114).

Similarly, if the polluters find themselves in a situation of perfect competition and the polluted parties form a single group, the polluted parties will receive

---

1) If the certificates held by the polluted parties correspond to a public good (quality of the air), it will sometimes be necessary to tax certificate sales in order to compensate other certificate holders for the general fall in quality resulting from the sale of certificates held by certain polluted parties (municipalities, but not individuals).

or will pay $C(p) - C(p_0)$ for the certificates. Thus the polluted parties will determine the optimum pollution level, will be the only ones to benefit from the optimisation gain and will cause the polluters to declare the exact cost of pollution abatement. Here, the certificate system corresponds to the principle of taxing the victims (see report at pages 87-114).

    d) In a market with perfect competition (large number of polluters and victims liable to purchase or sell certificates at the equilibrium price determined by the equivalence of supply and demand), the certificate price π is equivalent to the marginal cost at the optimum pollution level p^*.

$$\pi = \left[\frac{\partial D}{\partial p}\right]_{p=p^*} = \left[\frac{\partial C}{\partial p}\right]_{p=p^*}$$

    For reasons of competition, polluters and victims make exact declarations of costs. It the polluters initially hold certificates corresponding to the level p_0, they bear the cost $C(p^*) + p^* - p_0)\,\pi$ while the victims bear the cost $D(p^*)+(p_0-p^*)\pi$. Choosing the level p^* instead of the level p_0 minimise the costs of the two countries who divide the optimisation gain $T(p_0) - T(p^*)$. In conclusion, the certificate system becomes equivalent to the mutual compensation principle only when it is possible to set up a situation of perfect competition.

Exchange market for environmental rights

    If the polluting country owns or holds pollution rights corresponding to the reference level p_0 which is less than the optimum level p^*, it is then prepared to buy additional rights for a price less than $C(p_0) - C(p)$, for instance

$\alpha \left[C(p_0) - C(p)\right]$, and the polluted country is ready to sell rights for a price higher than $D(p) - D(p_0)$, say $\delta\left[D(p) - D(p_0)\right]$.

    If these transactions take place in an "exchange market" (1) by balancing supply and demand and effecting all transactions at the equilibrium price, the equilibrium will be situated at the level p_a, a solution of:

$$\alpha\,\frac{\partial C}{\partial p} + \delta\,\frac{\partial D}{\partial p} = 0$$

The polluting country will bear the net cost

$$T_M = C(p_a) + (p_a - p_0)\,\delta\,\frac{\partial D(p_a)}{\partial p}$$

and will choose the value of α which minimises this cost:

$$1 - \alpha = \frac{(p_a - p_0)\,\delta\,\dfrac{\partial^2 D(p_a)}{\partial p^2}}{\left|\dfrac{\partial C(p_a)}{\partial p}\right|} > 0$$

The polluted country will bear the net cost:

$$T_V = D(p_a) + (p_a - p_0)\,\alpha\,\frac{\partial C(p_a)}{\partial p}$$

and will choose the value of δ which minimises this cost T_V:

$$\delta - 1 = \frac{(p_a - p_0)\,\alpha\,\dfrac{\partial^2 C(p_a)}{\partial p^2}}{\dfrac{\partial D(p_a)}{\partial p}} > 0$$

---

1) This exchange market resembles the joint agency of the mutual compensation principle, but does not have its own budget.

Since $\alpha < 1$ and $\delta > 1$, the acceptable level p_a will be lower than the optimum level p^* and setting up an exchange market does not produce the optimum solution. On the other hand, if $p_o > p_a$, the acceptable level will be higher than the optimum level p^*.

The linear pollution charge

If the polluted country is able to choose the amount of compensation it will be offered by the polluting country in the event of substantial pollution, and which it will pay in the event of low pollution, it will not opt for an excessive amount for fear of having to pay a substantial bribe to the polluting country, nor a too low amount for fear of being victim of substantial pollution without compensation. The polluting country will choose the pollution level in accordance with the charge it will have to pay and its treatment costs.

The net costs in this system are:

$$\begin{cases} T_M = C(p) + t(p-p_o) \\ T_V = D(p) + t(p_o-p) \end{cases}$$

where p_o is the reference or exemption level.

The polluting country has to pay a charge t and choses a pollution level p_a with a view to minimising its net cost T_M. Since in this way it declares, to the polluted country, the marginal treatment cost, the polluting country can choose an inaccuracy factor α for the treatment cost such that the level p_a is given by

$$t + \alpha \frac{\partial C}{\partial p} = 0 \tag{3}$$

The polluted country chooses the amount t and an inaccuracy factor δ for the damage cost in order to minimise its net cost T_V. The amount t is given by

$$\delta \frac{\partial D}{\partial p_a} - t \frac{\partial p_a}{\partial t} + (p_o - p_a) = 0 \tag{4}$$

and the factor δ by

$$\frac{\partial D}{\partial p_a} \frac{\partial p_a}{\partial \delta} + \frac{\partial t}{\partial \delta}(p_o - p_a) - \frac{\partial p_a}{\partial \delta} t = 0 \tag{5}$$

Combining equations (4) and (5), we have

$$\frac{\partial D}{\partial p_a}(1 - \delta) = 0$$

since $\dfrac{\partial p_a}{\partial t} = \dfrac{\partial p_a}{\partial \delta} \Big/ \dfrac{\partial t}{\partial \delta}$

as p_a, t and α are linked through equation 3 (1).

The polluted country will consequently declare the exact damage cost ($\delta = 1$).

---

1) From equation 3, it is seen that

$$\frac{\partial t}{\partial \alpha} = -\left[\alpha \frac{\partial^2 C}{\partial p_a^2} \frac{\partial p_a}{\partial \alpha} + \frac{\partial C}{\partial p_a}\right]$$

$$\frac{\partial t}{\partial \delta} = -\alpha \frac{\partial^2 C}{\partial p_a^2} \frac{\partial p_a}{\partial \delta}$$

$$\text{and } \frac{\partial p_a}{\partial t} = -\frac{1}{\alpha \dfrac{\partial^2 C}{\partial p_a^2}}$$

If equations (3) and (4) are combined, we have

$$\alpha \frac{\partial C}{\partial P_a} + \frac{\partial D}{\partial P_a} = \frac{p - P_0}{\frac{\partial P_a}{\partial t}} \tag{6}$$

where $\frac{\partial P_a}{\partial t} < 0.$

If $\alpha \leqslant 1$ and $p_0 \leqslant p_a$, or if $\alpha \geqslant 1$ and $p_0 \geqslant p_a$, the acceptable pollution level is always between the level p_0 and the optimum level.

The net cost T_M is minimum with respect to α if

$$\frac{\partial C}{\partial P_a} \frac{\partial P_a}{\partial \alpha} + \frac{\partial t}{\partial \alpha} (p_a - p_0) + t \frac{\partial P_a}{\partial \alpha} = 0 \tag{7}$$

By combining equations (4) and (7) and taking account of the fact that $\delta = 1$, we have

$$\left[\frac{\partial C}{\partial P_a} + \frac{\partial D}{\partial P_a} \right] = \frac{\partial C}{\partial P_a} \frac{P_a - P_0}{\frac{\partial P_a}{\partial \alpha}} \neq 0$$

Generally speaking, the acceptable pollution level differs from the optimum level. If $\frac{\partial P_a}{\partial \alpha} > 0$, it can easily be shown that the acceptable pollution level is between the reference level p_0 and the optimum level $p^{\ast}$. It is in the interests of the polluting country to choose the coefficient α defined by the formula

$$(1 - \alpha) = \left[p_a - p_0 \right] \left[\frac{1}{\frac{\partial P_a}{\partial \alpha}} - \frac{1}{\frac{\partial C}{\partial P_a} \frac{\partial P_a}{\partial t}} \right].$$

obtained from equations (6) and (7).

If $p_0 > p_a$, the polluting country has an interest in providing a value as high as possible for $\left| \frac{\partial P_a}{\partial t} \right|$ in order to get a bribe as high as possible. If $p_0 < p_a$, this country finds an interest in choosing a small value for $\left| \frac{\partial P_a}{\partial t} \right|$ in order to bear a small rate for the charge.

$$t = \frac{\partial D}{\partial P_a} + \frac{P_a - P_0}{\frac{\partial P}{\partial t}} < \frac{\partial D}{\partial P_a} \qquad \text{if } p_0 > p_a$$

For this situation to be reached, the polluting country would have to act as if the treatment cost was different from its real value when various rates apply. This fairly complex strategy will be effective for the polluting country only if the additional cost it bears due to not declaring its real costs is less than the profit gained by so acting. This strategy will succeed if the polluted country is satisfied with only limited information for determining the treatment cost, i.e. if the polluted country believes that the polluting country is not trying to conceal the true treatment cost and if the polluted country does not change the rate of the charge too frequently.

If the reference level p_0 is close to the optimum, the taxation system produces a satisfactory result and the transactions take place virtually at the ideal unit cost. However, it is difficult to choose the level p_0 close to the optimum level since this optimum level is unknown when the reference level p_0 is being chosen.

The system of the linear charge is intended to enable the polluted country to choose the rate of compensation as it wishes and thus to weigh the advantages and disadvantages of this choice. If the polluting country were to agree to declare the exact treatment cost ($\alpha = 1$), this system would produce a non-optimum solution, (except if $p = p_0$ or $\frac{\partial^2 C}{\partial p^2} = 0$, and it would be preferable to use the modified victim-pays principle.

Fixed shared responsibility (1)

If the polluting country bears the fixed fraction f $(0 < f < 1)$ of the total cost, and the polluted country the remaining fraction $(1-f)$, the net cost borne by the polluting and polluted countries which declare costs α C and δ D are:

$$\begin{cases} T_M = C + f \delta D + (f - 1) \alpha C = (1 - \alpha + \alpha f) C + f \delta D \\ T_V = D - f \delta D - (f - 1) \alpha C = (\alpha - \alpha f) C + (1 - f \delta) D \end{cases}$$

If the polluting country chooses the pollution level p_a which minimises the sum of the declared costs, the level p_a is a solution of

$$\alpha \frac{\partial C}{\partial p} + \delta \frac{\partial P}{\partial P} = 0$$

This is a rational choice insofar as the polluting country does not wish to reveal that it has provided a wrong estimate of the treatment cost.

The net costs borne by the polluting and polluted countries are minimal if

$$\frac{\partial T_M}{\partial \alpha} = 0 \text{ and } \frac{\partial T_V}{\delta \partial} = 0, \text{ i.e. if } \alpha \text{ is a solution of}$$

$$\alpha - 1 = \frac{(1 - f)}{\frac{\partial P_a}{\partial \alpha}} \frac{C(p_a)}{\left| \frac{\partial C(p_a)}{\partial p} \right|}$$

and if δ is a solution of

$$\delta - 1 = \frac{f \ D(p_a)}{\left| \frac{\partial P_a}{\partial \alpha} \right| \frac{\partial D(p_a)}{\partial p}}$$

It is immediately clear that it is in the interest of both countries to exaggerate the costs ($\alpha > 1$ and $\delta > 1$). Moreover, the optimum level will not be reached since the above equations lead to

$$\frac{\alpha - 1}{\alpha^2} \Big/ \frac{\delta - 1}{\delta^2} = \frac{1 - f}{f} \frac{C(p_a)}{D(p_a)}$$

and because the fraction f is chosen before the actual or declared costs are known, so that $\frac{1 - f}{f} \frac{C(p_a)}{D(p_a)} \neq 1$ and $\frac{1 - f}{f} \frac{\alpha C(p_a)}{\delta D(p_a)} \neq 1$. Nevertheless, the divergence from the optimum solution could be small if the factor f were close to $\frac{C(p^{\star})}{C(p^{\star}) + D(p^{\star})}$ i.e. if the factor f were chosen so as to give approximately the same result as the one which would be obtained using the polluter-pays principle in the strict sense.

Modified shared responsibility

The system of shared responsibility where the fraction f depends on cost estimates at the acceptable level has already been investigated (Paper by H. Smets, Annex I, section e.3, Problems of Transfrontier Pollution, OECD, 1974). The following two cases are of special interest.

$$f = \frac{\alpha C(p_a)}{\alpha C(p_a) + \delta D(p_a)} \quad \text{or} \quad \frac{1 - f}{f} \frac{\alpha C(p_a)}{\delta D(p_a)} = 1$$

1) See paper by H. Smets given in "Problems of Transfrontier Pollution", OECD, 1974. A fraction f is chosen instead of a reference level p_o for which the polluter bears the treatment cost.

The net costs for the polluting country and the polluted country are $T_M = C(p_a)$ and $T_V = D(p\ )$ where p_a is the level which minimises the total declared cost $[\alpha\, C(p_a) + \delta\, D(p_a)]$. This system is therefore equivalent to the conventional polluter-pays principle examined above: the factors α and δ cannot be specified but they could be very high.

$$f = \frac{\delta\, D(pa)}{\alpha\, C(pa)\ +\ \delta\, D(pa)}$$

The net costs for the polluting and polluted countries here are

$$\begin{cases} T_M = \delta D(p_a) - (\alpha - 1)\, C(p_a) \\ T_V = \alpha\, C(p_a) - (\delta - 1)\, D(p_a) \end{cases}$$

where p_a is the pollution level which minimises the total declared cost $\left[\alpha\, C + \delta\, D\right]$

The net costs are minimal if $\dfrac{\partial T_M}{\partial \alpha} = 0$ and $\dfrac{\partial T_V}{\partial \delta} = 0$

i.e. if

$$2\,\alpha - 1 = \frac{C(p_a)}{\dfrac{\partial C(pa)}{\partial p}\ \left|\ \dfrac{\partial Pa}{\partial \alpha}\ \right|} > 0$$

$$2\,\delta - 1 = \frac{D(p_a)}{\dfrac{\partial D(pa)}{\partial p}\ \dfrac{\partial Pa}{\partial \delta}} > 0$$

The solution is non-optimal because

$$\frac{2\,\alpha - 1}{\alpha} \bigg/ \frac{2\,\delta - 1}{\delta} = \frac{\alpha\, C(pa)}{\delta\, D(pa)}$$

and because there are no special reasons why

$$\frac{\alpha\, C(pa)}{\delta\, D(pa)} \quad \text{or} \quad \frac{C(pa)}{D(pa)}$$ should be equal to unity. Under this system, the polluter has to bear the residual damage cost and the polluted country the cost of pollution control (Reversed Polluter Pays principle).

In comparison with the polluter-pays principle, it has the advantage of not generating too inaccurate cost estimates. Like the polluter-pays principle, it calls for no prior information (exemption, reference level or fixed fraction of responsibility f) and it is equally "logical" from the equity point of view as the polluter-pays principle in the strict sense. However, this method does have the disadvantage of giving different results from those invoking the polluter-pays principle, which is generally recognised.

B. THREE-PARTY SYSTEMS

Three-party systems are systems in which an agency external to the two parties is made responsible for selling or buying "certificates" for "rights" in respect of pollution or treatment, to collect charges or taxes or to make bribes. This agency does not need to have a balanced budget and cannot be owned by the two parties concerned in the pollution (polluting and polluted parties) otherwise it is equivalent to the systems examined in Section A. The following is an examination of different systems where neither the treatment nor the damage cost is well known.

The so-called "Swiss corporation"

If the two countries assign to an agency the power to manage the transfrontier environment (or some common resource) against two payments to compensate them for giving up all their environmental rights, and if this agency is empowered to sell "pollution rights" to potential polluting countries and "treatment rights" to those seeking a clean environment, and if this agency is able to withhold these rights (1) in order to obtain the maximum price by virtue of its monopolistic position, then the agency will be able to exact the price $C(p_0) - C(p)$ from the polluting country holding an initial right p_0 and wishing to pollute up to the level $p > p_0$. At the same time, the agency may sell to the polluted country the service corresponding to a pollution level p which is less than p_{max} for a price $D(p_{max}) - D(p)$. The joint agency will be able to obtain this high price insofar as it can actually threaten the victim with a high pollution level p_{max} (for example by not carrying out any treatment).

Since the net costs for the polluting and polluted countries are constant $[C(p_0)$ and $D(p_{max})]$, the two countries will be prepared to agree that the pollution level chosen be the one which maximises the agency's profit

$$D(p_{max}) - D(p) + C(p_0) - C(p)$$

i.e. the optimal level p^*. However, this agreement is linked to the idea that the two countries will share the profit of this agency, which A.D. Scott calls the "Swiss Corporation", for example in the form of a lump sum transfer paid by the corporation to the two countries (2).

The Swiss Corporation does not know what costs are borne by the two countries, and in the context of international relations can only take note of the fact that the polluting country is not prepared to pay more than $\alpha [C(p_0) - C(p)]$ to buy the right to pollute to the level p. Similarly, the polluted country does not wish to pay more than $\delta [D(p_{max}) - D(p)]$ to benefit from a pollution level $p < p_{max}$. In this case, the corporation will have to choose a pollution level p_a which minimises $\alpha C(p) + \delta D(p)$ instead of the optimal level p^*. The costs borne by the polluting country and by the polluted country

$$\begin{cases} T_M = C(p_a) + \alpha [C(p_0) - C(p_a)] \\ T_V = D(p_a) + \delta [D(p_{max}) - D(p_a)] \end{cases}$$

are minimal if α and δ are given by

$$1 - \alpha = \cfrac{C(p_0) - C(p_a)}{\left|\cfrac{\partial C(p_a)}{\partial p} \,\middle|\, \cfrac{\partial p_a}{\partial \alpha}\right|} > 0$$

$$1 - \delta = \cfrac{D(p_{max}) - D(p_a)}{\cfrac{\partial D(p_a)}{\partial p} \,\middle|\, \cfrac{\partial p_a}{\partial \delta}} > 0$$

---

1) The polluter may not pollute to a greater extent than that corresponding to the pollution rights purchased and the victim is unable to benefit from an environmental quality higher than the one he has bought.

2) See A.D. Scott in "Problems of environmental economics", OECD, 1972. It should be noted that if the "Swiss corporation" were to divide its profit between the two countries, the net costs for each country would be :

$$\begin{cases} T_M = C(p_a) + f\, \delta \left[D(p_a) - D(p_{max})\right] + (f - 1)\, \alpha \left[C(p_a) - C(p_0)\right] \\ T_V = D(p_a) - f\, \delta \left[D(p_a) - D(p_{max})\right] - (f - 1)\, \alpha \left[C(p_a) - C(p_0)\right] \end{cases}$$

which would be equivalent to the system of fixed shared responsibility (negotiation between two parties).

so long as the countries have to purchase rights from the corporation, they will tend to underestimate the costs in order to reduce the profit of the corporation which they do not own. The above equations lead to the expression

$$\frac{1 - \alpha}{\alpha\, 2} \left/ \frac{1 - \delta}{\delta\, 2} \right. = \frac{C(p_o) - C(p_a)}{D(p_{max}) - D(p_a)}$$

The acceptable level of pollution will differ from the optimal level since there is no special reason why the prices paid to the corporation by the countries should be the same.

Compulsory public auction (1)

If an agency sells "pollution rights" and is obliged to sell all the available rights to the highest bidder, the polluting country's highest bid for each right will be the marginal treatment cost, taking account of the number of rights it holds, and the polluted country will bid at least the marginal damage cost for each right, in view of the number of rights acquired by the polluted country. If the polluting country holds rights corresponding to p_o, the polluted country bids $\delta\, \dfrac{\partial D(p_o)}{\partial p}$ for a right and the polluting country bids

$$\left| \alpha\, \frac{\partial C(p_o)}{\partial p} \right| \quad \text{for this right.}$$

The polluting country will purchase all the rights until

$$\alpha \left| \frac{\partial C}{\partial p} \right| = \delta\, \frac{\partial D}{\partial p}$$

since each right is worth more to it than to the polluted country. If the polluted country does not outbid the polluting country when it does not intend to purchase (benevolent behaviour), the polluting country will pay $\delta\, D(p)$ for the rights acquired up to the level p_a. Beyond the level p_a, the polluting country will bid

$$\delta\, \frac{\partial D(p_a)}{\partial p} = \alpha \left| \frac{\partial C(p_a)}{\partial p} \right|$$

for each offer, since a right acquired at any time during the auction allows the polluting country to avoid that treatment cost and the polluted country will agree to pay more for these rights since

$$\delta\, \frac{\partial D(p)}{\partial p} > \alpha \left| \frac{\partial C(p)}{\partial p} \right| \quad \text{for } p > p_a,$$

Finally, the polluted country will pay

$$(p_{max} - p_a)\, \alpha \left| \frac{\partial C(p_a)}{\partial p} \right|$$

for the additional rights and will bear the cost $D(p_a)$. The polluting country will bear the cost $T_M = C(p_a) + \delta\, D(p_a)$. The selling agency will collect

$$D(p_a) + (p_{max} - p_a) \left| \frac{\partial C(p_a)}{\partial p} \right|$$

There is no point in the polluting country declaring a treatment cost different from reality since the net cost is minimal for $\alpha = 1$. It is in the interest of the polluted country to proceed as if the damage cost were lower than it really is since its net cost is minimal for

---

1) See D. R. Lee and C. W. Howe: "A new look at pollution rights", preprint, Department of Economics, University of Colorado, 1973.

$$1 - \delta = \frac{(p_{max} - p_a) \dfrac{\partial^2 C(p_a)}{\partial p2}}{\dfrac{\partial D(p_a)}{\partial p}} > 0$$

Thus the system described does not lead to economic efficiency because the polluted country is liable to underestimate the damage cost, and the acceptable pollution level is then higher than the optimal.

For the system to operate as described, the agency has to auction many more rights than those which correspond to the optimal level. If the polluting country initially holds more rights than the optimal level, the system reverts to the sale of certificates by the polluting country to the polluted country and no auction is necessary. If the damage cost is well known by the two countries ($\delta = 1$), it is not necessary to introduce the system of public auction and a system of pollution charges (modified civil liability principle) suffices.

Modified public auction

If the auction system is modified so that the residual rights corresponding to $(p_{max} - p_a)$ are sold in a block once the polluted country has purchased a right, the polluting country can bid up to $\alpha \left[C(p_a) - C(p_{max}) \right]$ for these rights and the polluted country may purchase these rights for the same amount. The costs borne by the polluting country and the polluted country are

$$\begin{cases} T_M = C(p_a) + \delta\, D(p_a) \\ T_V = D(p_a) + \alpha \left[C(p_a) - C(p_{max}) \right] \end{cases} \text{ where } p_a \text{ is a solution of}$$

$$\alpha \frac{\partial C}{\partial p} + \delta \frac{\partial D}{\partial p} = 0$$

The polluting country will be obliged to buy rights up to the level p_a to conceal the fact that its declared estimate of treatment cost may be erroneous, although its net cost is minimal for a pollution level below the level p_a. The polluting country will see at once that in order to minimise its own net cost it should declare the exact value of the treatment cost ($\alpha = 1$). The same will apply to the polluted country which will choose the value $\delta = 1$.

The modified public auction system leads to an efficient solution. The selling agency will collect

$$D(p_a) + C(p_a) - C(p_{max})$$

and will therefore have a minimal income at the optimal level sought. Hence this agency will not act as a profit-making body but as an international public service. Any deficit it incurs will have to be financed without affecting its independence from the two countries concerned.

The modified public auction system has the same structure as that based on the mutual compensation system, but has a different institutional form. Combined pollution tax and treatment tax might however be more easily accepted than "rights" negotiable on the market.

Practical considerations

a) The system of public auction should not be used every year since direct negotiation between the two countries leads to an efficient solution when the exchanges take place close to the optimal pollution level (for example by the sale of certificates). However, it might be asked whether auctions should not be organised when appreciable changes occur in economic and technological situations.

b) Pollution permits should not be auctioned too frequently in order to avoid introducing too much uncertainty in investment decisions.

c) If an economic transactor wishes to purchase or sell certificates after the auction, this might be made conditional on the transaction being public and on certificate holders (current polluters and pollutees) having the right to acquire certificates at the same price or by making a higher bid (a price maintenance mechanism to avoid rights being allocated to an enterprise as an installation grant).

d) Difficulties may arise in the case where an influential polluting enterprise buys all pollution rights in order to compel the other, weaker polluting enterprises to cease their activities.

C. COMPARISON OF DIFFERENT GAME RULES

It is possible to compare the different game rules if the strategies followed by the two parties are specified. It is assumed here that the parties choose simple strategies, i.e. freely choose the uncertainty factors α and δ affecting the treatment and damage costs. Moreover, it is assumed that the parties have agreed that each will bear a fraction of the total cost. Specifically, the polluter agrees to bear the treatment cost so as to reduce pollution to the level p_0 at his expense and the victim agrees to bear the residual cost.

Model

The treatment cost is assumed to be $\frac{1}{2}(2-p)^2$ and the damage cost $\frac{1}{2}p^2$. The reference level is 1.5 while the optimal level is 1. The polluter agrees to bear up to the corresponding $\frac{1}{8}$ of the treatment cost and the victim agrees to bear up to the residual cost $\frac{9}{8}$. The total cost of the reference level is $\frac{10}{8}$ which exceeds by 25 per cent the minimum cost of the optimal level ($\frac{8}{8}$).

Results

1. Conventional polluter-pays principle

   No result possible.

2. Modified victim-pays principle ($p_0 = 1.5$)

   $\alpha = 1.72$ $\delta = 1$ $p_a = 1.27$

3. Modified civil liability ($p_0 = 1.5$)

   $\alpha = 1$ $\delta = 0.62$ $p_a = 1.23$

4. Exchange market for rights ($p_0 = 1.5$)

   $\alpha = 1.25$ $\delta = 0.75$ $p_a = 1.25$

5. Linear charge ($p_0 = 1.5$)

   $\alpha = 1$ $\delta = 1$ $p_a = 1.16$

   Note: The amount of the charge is 0.835 while the marginal damage cost at the optimum is 1.

6. Shared responsibility ($f = 0.1$)

   $\alpha = 2.8$ $\delta = 1.2$ $p_a = 1.4$

7. Swiss corporation ($p_{max} = p_0 = 1.5$)

   $\alpha = 1.46$ $\delta = 0.71$ $p_a = 1.33$

8. Mutual compensation (and modified public auction) ($p_0 = 1.5$)

   $\alpha = 1$ $\delta = 1$ $p_a = 1$

Other systems

The following results correspond to different initial conditions from those considered above and which cannot be achieved with the following systems:

160

9. Reversed polluter-pays principle (modified shared responsibility) (p_o = ?)

    $\alpha = 1$ $\delta = 1$ $p_a = 1$

    Note: This correct result is due to the special model which is completely
          symmetrical. Generally speaking, the reversed polluter-pays
          principle is less satisfactory than the mutual compensation prin-
          ciple.

10. Compulsory public auction (p_{max} = 2)

    $\alpha = 1$ $\delta = 0.5$ $p_a = 1.33$

Conclusion

These calculations show to what extent the costs estimates are likely to be
inaccurate and indicates the influence of these errors on the differences from the
optimal pollution level. From the practical standpoint, it may be justifiable to
choose a system different from the mutual compensation principle if sufficient
information is available on the costs functions and if, in the case in question,
errors in the cost estimates are insignificant.

APPLICATION OF THE MUTUAL COMPENSATION PRINCIPLE
TO A SPECIFIC, AND HYPOTHETICAL CASE

A. FIRST CASE

An Imaginary Treaty

Let us take the case of two countries lying along a river polluted by the upstream country. A treaty is concluded by the countries by virtue of which the maximum acceptable pollution level p_o, the cheapest method of pollution control is defined, and its cost $C(p)$ can be calculated in a way satisfactory to both parties. The upstream country agrees to bear a proportion x of the cost of pollution control $C(p_o)$ when the pollution level is p_o, while the downstream country bears the proportion y. In compensation for the transfer $yC(p_o)$, the upstream country undertakes not to exceed the pollution level p_o.

Analysis of the Treaty

The two countries have agreed on the cost of pollution control but not on the cost of damage. The pollution level p_o is of course the "optimum" level if the downstream country makes rational decisions and has perfect information and if it does not foresee future changes in the accepted equity principles. Thus the downstream country, on the basis of its knowledge of the damage, has not offered to pay the additional cost $C(p_1) - C(p_o)$ in order to achieve the lower level p_1, and has not accepted a higher pollution level p_2 in exchange for a reduction in its contribution by the amount $C(p_o) - C(p_2)$.

There is no economic incentive for the upstream country to do everything it can to maintain the pollution level below the value p_o. If a pollution episode occurs, the downstream country will probably not be compensated, since the agreement is silent on this point. However, the marginal damage cost is relatively simple to evaluate roughly since it equals the marginal cost of pollution control which is known (equivalence of marginal costs at the optimum level). If the upstream country wants to raise the level p_o, it will have to offer the downstream country the additional damage cost $D(p) - D(p_o)$, which it will be able to estimate from the marginal cost of pollution control. Yet nothing will compel the downstream country to agree to this request, since this country can point out that the cost of damage increases very rapidly beyond the level p_o (for psychological reasons, this is what is most likely to happen). If the downstream country wants further treatment, it will have to pay the additional treatment cost to the upstream country. However, the upstream country could also say that the cost of antipollution grows very rapidly for decreased pollution ($p < p_o$).

To sum matters up, the agreement described lacks a clause dealing with the liability of the polluting country in the event that the pollution level is exceeded, and a clause stating the conditions under which the countries can be allowed to change the pollution level p_o. To get round this difficulty, the treaty would contain a provision for termination at any time on short notice. Then it could be feared that countries would not implement antipollution measures which require significant capital outlays.

Use of the Mutual Compensation Principle

If the countries decide to apply the mutual compensation principle, they would set up a joint agency to which an annual declaration (index t) is submitted concerning the costs of pollution control (1) and of damage $[C_t(p)$ and $D_t(p)]$. The upstream country pays the tax $[D_t(p) - D(p_o)]$ into the joint agency and bears the cost $[C_t(p) - yC(p_o)]$ while the downstream country pays the tax $[C_t(p) - C(p_o)]$ and bears the cost $[D_t(p) + yC(p_o)]$. The two countries would agree that the joint agency shall fix the pollution level (treatment in the polluting country) by determining the level p^* which minimises the total amount of taxes levied.

$$\left[C_t(p) - C(p_o) + D_t(p) - D(p_o) \right]$$

The upstream country undertakes to reduce the release of waste so as not to exceed the level p^* determined by the agency. If the pollution level is exceeded, the upstream country would pay a higher tax and the downstream country would receive an indemnity (Annex 1). The level p^* is intended to vary with time in order to remain optimal in spite of cost changes. The countries pay or receive compensation to adjust for the change from the initial situation p_o. The mutual compensation principle establishes necessary conditions for the two countries to maintain economic efficiency through time by means of transactions at the unbiased cost concerning pollution rights or treatment rights which they possess.

The transfers E_1 and E_2 of the system described in Annex 1 are given in this case by:

$$\begin{cases} E_1 = yC(p_o) + D(p_o) \\ E_2 = xC(p_o) \end{cases}$$

Since both countries know the cost of pollution control $C(p_o)$, the transfer E_1 is quite independent of the factor α relating to the treatment cost, and the transfer E_2 quite independent of the factor δ relating to the damage cost. Moreover, the level p_o is independent of these factors α and δ since it corresponds to the optimum level in the initial year.

Since the budget of the joint agency is not balanced, it will fluctuate through time. If the upstream country becomes industrialised $[C_t(p_o) > C(p_o)]$ and the downstream country seeks better quality $[D_t(p_o) > D(p_o)]$, the funds of the joint agency will rise. If these funds exceed those assigned to the projects of common interest (management of the treaty, pollution surveillance, research, etc.) and reach a substantial level, the countries may agree to amend the treaty in view of the considerably changed situation.

The transfers E_1 and E_2 would then be redefined - for instance by so altering the fractions x and y - as to reflect the new situation of the countries in relation to each other and in order to reduce the total amount of taxes paid to the joint agency. The accumulated sum of taxes could be allocated to the joint agency as an initial grant.

When the changes in the values $C_t(p)$ and $D_t(p)$ are caused by changes in the economic situation for which the causality principle (2) can be employed, the country which does not change will not have to bear any increase in costs. If the upstream country creates polluting industries and hence causes an increase in the optimal level of discharges

$$C_t(p) > C_{t-1}(p), \; D_t(p) = D_{t-1}(p), \; p_t^* > p_{t-1}^* \,,$$

the causality principle requires the joint agency to compensate the downstream country by paying it in this case

$$C_t(p_t^*) - C_{t-1}(p_{t-1}^*) + D_t(p_t^*) - D_{t-1}(p_{t-1}^*)$$

---

1) The pollution control costs $C_t(p)$ vary because of technical innovation and industrialisation.

2) "He who changes, pays", see report at pages 31-54 (e.g. creation of a new polluting activity in the upstream country or a new pollution sensitive activity in the downstream country).

(which is equivalent to cancelling the change in the downstream country's tax, and compensating that country for the additional damage caused by industrialisation of the upstream country). When the causality principle is invoked for a particular change, the budget of the joint agency it not affected and the period of validity of the treaty is not reduced.

Therefore the mutual compensation principle corrects some imperfections in the treaty initially described by covering the case in which the pollution level is exceeded and the case where **revision** of the initially accepted level is requested. It enables one to maintain economic efficiency through time and therefore offers financial advantages to the countries. This result is obtained because the mutual compensation principle gives the countries every reason to believe that the estimates declared each year match the facts. Hence, the downstream country will not have to wonder why the cost of pollution control changes, after having accepted its initial evaluation, nor will the upstream country have to worry about any rapid increase in damage above certain pollution levels. Thus the two countries will reduce the cost of obtaining information regarding costs occurring in the other country. They will however incur information costs in ascertaining the cost they have to bear in the first instance, so as to avoid the negative effects of using bad estimates of such cost.

Moreover, the mutual compensation principle should be such that a treaty need not constantly be challenged, since it is flexible in terms of time. Yet it can include an automatic cancellation clause, which will take effect when the accumulated imbalance of the joint agency's budget has become substantial (objectively showing that the initial situation has considerably changed and that a new negotiation has become necessary).

B. CASE TWO

Description: a new source of pollution (1)

A country is considering siting a polluting enterprise (2) close to its frontier with another country; the enterprise is able to cause damage in the other country as a result of transfrontier pollution. The countries agree on the principle of equity according to which the polluted country will bear the damage cost when the pollution level is below the level p_0 (which may be very low), and the polluter country will bear the entire cost due to transfrontier pollution less the residual damage cost at level p_0 when the pollution level is greater than p_0.

The problem is to define the optimum pollution level that will satisfy the interests of the two countries (3) in circumstances when they know the cost they will bear in the first place (damage or pollution control) but have not reached agreement on what these costs should be. In this type of problem, allowance will be made for the highly subjective nature of damage at the pollution levels considered. The countries realise that the reference level p_0 is excessively low,

---

1) This study does not go into the symmetrical case where an area of reduced pollution is established in the downstream country. However the principles described apply equally.

2) Examples: a refinery, a nuclear power station or a chemical plant which is responsible for mild disamenity or a slight change in death rate, together with a more or less troublesome psychological annoyance which however in normal circumstances does not produce evident damage to human health. Such psychological annoyance includes disamenities and effects which are unfavourable to the development of tourism. It is assumed that negotiations are related to a pollution parameter and not to the principle of setting up a more or less polluting enterprise.

3) This problem would lose much of its significance if the polluting enterprise were to produce advantages and disadvantages equivalent in the two countries. In theory therefore, it might be agreed that revenue from local taxes should be distributed amongst the most seriously affected communities in the two countries, instead of being allocated solely to the community where the enterprise is sited, and that the two countries would enjoy equivalent economic advantages from the enterprise (e. g. shared ownership and electricity or oil production) just as if the enterprise were located astride the border.

thus leading to substantial expenditure on pollution control, but realise also that choosing a higher reference level would result in extremely unfavourable reactions on the part of the exposed population. In order to reach a more satisfactory solution, it would probably be preferable to find some way of getting a pollution level higher than p_0 accepted freely by the exposed population in the polluted country.

Proposed solution

Countries which had accepted the level p_0 and the equity principle, "he who changes, pays", encounter difficulties in negotiation because of substantial disagreement over the costs of pollution control and of damage. To resolve this problem, they decide to invoke the mutual compensation principle, and agree that the polluted region should be paid financial compensation if it agrees to accept a pollution level higher than p_0. The net costs for the polluting and polluted countries then become:

$$\begin{cases} T_M = C(p) + \delta D(p) \\ T_V = D(p) + \alpha \left[C(p) - C(p_0) \right] \end{cases}$$

when the polluting country pays the pollution tax $\delta D(p)$ to a joint agency and the polluted country is paid compensation $\alpha \, [C(p_0) - C(p)]$ by that agency. The budget of the joint agency is:

$$\delta D(p) + \alpha \, [C(p) - C(p_0)]$$

If the two countries are benevolent (good international relations), the polluting and polluted countries will declare the actual costs they will bear in the first place (mutual compensation principle). Moreover, there is no point in the polluting country giving an incorrect figure for the treatment cost at the level $C(p_0)$ because it would not benefit by doing so. Hence the optimum pollution level p^* will be reached and economic efficiency guaranteed.

Both the polluting and polluted countries will bear lower costs then they would at the level p_0 [equal gain for each country:

$$C(p_0) + D(p_0) - C(p^*)]$$

and both benefit from recourse to this solution. The budget of the joint agency will be in deficit if the level p_0 is very low, and will show a surplus if the level p_0 is close (1) to the optimum level p^*. The maximum value for the joint agency's budget is $D(p_0)$, which is positive although probably fairly low. If the agency's budget is in deficit, the net cost borne by the polluted country is negative, and that country is thus better off economically as a result of transfrontier pollution (the polluted country receives a compensation payment which is higher than the cost of the damage suffered).

A solution to reduce the unbalanced budget of the joint agency would be for the States to cover this deficit. To the extent that the respective Governments abstain from influencing the parties involved in the estimation of costs (polluting enterprise and polluted region), the desired result would be achieved.

Discussion

Area of negotiation

The negotiations will cover a limited range of pollution levels. An absolutely zero pollution level (2) is excluded as unrealistic and too costly. It reflects a too inflexible attitude which is usually inapplicable on an international level. High

---

1) In this case it would not appear very realistic for the level p_0 to be close to the optimum level, because p_0 has to be chosen before the costs are known. These costs change with time, and the level p_0 is usually much lower than the optimum level.

2) The same applies to a zero increase of pollution above its natural level.

pollution levels are also excluded since they would never be tolerated by the polluting country (it would be unthinkable to build an installation that would seriously threaten the health of any population whatsoever). Within these limits, there are causes of annoyance (disamenities) and hazards to health of a statistical nature (rise in the death rate, risk of accident, and so on).

The principle of compensation

With the proposed solution, the polluted country is offered compensation in respect of new pollution. It is no longer forced to accept environmental damage under a variety of pretexts, but freely participates in the choice of a reasonable pollution level, while making the most of the advantages arising from the new socio-economic situation. With this system, the economic situation of the polluted country usually improves when the polluting enterprise is established although the level of polluting is higher. This solution appears especially attractive for the polluted country, which has to bear no significant additional cost and has its own environmental rights recognised.

The polluting country will tend rather to reject this solution, preferring to build the installation with a maximum pollution level set close to the optimum level p^*, and refusing to pay a pollution tax. In this way the polluting country does not have to acknowledge that damage exists, and can thus claim that it does not.

This attitude on the part of the polluter can no doubt be justified on a national scale where the economic agents are collectively responsible and actually experience the advantages and disadvantages of their activities. The government may, in the general interest, take steps which are injurious to certain parties and grant compensation only when the damage is excessive. For instance, only a few governments have compensated people living near a new motorway or airport because of noise.

This attitude is less easy to justify at an international level, because the economic transactors - both polluters and polluted - at both sides of the frontier do not feel jointly involved. Quite often, the polluted parties do not obtain any real advantage from the actions of foreign polluters (e.g., as local taxes, industrial development, new jobs or more favourable prices). This situation is particularly evident when the polluted parties had refused the polluting enterprise now built just across the frontier to be set up on their own side, or have decided to forbid the establishment on their side of any such polluting enterprises.

To reach a situation of collective solidarity such as exists at national level, transfers would have to be arranged between polluters and pollutees. The pollution tax is one way of doing this, and induces the polluting country not to decide to establish polluting enterprises at its boundaries in order to increase its national welfare at the expense of the well-being of the neighbouring country.

Hence this solution respects the sovereignty of the polluting country, whose polluting enterprises are not prohibited, and the sovereignty of the polluted country, which escapes having any pollution level imposed on it. The polluting country gains the right to pollute up to a certain limit subject to payment of a tax, and the polluted country the right to maintain its economic situation by means of a compensatory payment. The conflict of interest is resolved through a binding deal in which each country is free to declare a price and gains equal advantage.

The objection will be raised that the proposed solution opens the door to excessive demands for compensation. There is no basis for this argument with the proposed solution because the polluted country (so long as it remains benevolent) will not declare damage different from reality, nor will there be any point in its mounting a campaign of opposition (1) since it will be compensated for its economic situation to remain unchanged.

---

1) The situation is different if the polluted country does not receive compensation, if it is subjected to a pollution level which falls as objections become more vigorous, or if it is paid compensation in relation to the intensity of its opposition (the Polluter-Pays principle in the strict sense, or the Civil Liability principle, see Annex 2). Such systematic oppositions are the consequence of the polluter's decision not to act in such a way that pollutees do not stand a loss of well-being.

The compensation offered to the polluted country throws light on the true extent of the damage, by inducing the polluted parties to assess matters more reasonably, to weigh the advantages and disadvantages of further pollution and to adopt a less extreme attitude. Thus compensation has the result of restricting damage assessments and drawing attention to problems of pollution before it is too late.

Calculating the pollution tax

The pollution tax is bound up with the damage cost as perceived by the polluted parties (and not only to the monetary damage cost as it would be worked out by some legal body). The damage cost may be lower if information is inaccurate, misleading or untruthful and if the public is not aware of the hazards to which it is exposed. Conversely, the damage cost will be higher if the public develops feelings of antagonism, anxiety or aversion to the risk. The damage cost used here is that perceived by the polluted country so that the decision taken corresponds not to a technocratic view of the damage but to the real socio-economic situation in the polluted country. The estimation of this cost will introduce all the psychological and political aspects of the problem; it will not be carried out by experts or by an international organisation whose competence would not be easily acceptable to countries anxious to safeguard their sovereignty.

This cost will probably always be higher than that which would be perceived in similar circumstances by pollutees within the polluting country, since they are able to obtain indirect advantages from the polluting enterprise (for instance, the trading licence, or other local tax). This difference in perception should not be interpreted as xenophobia applied to pollution. The damage cost will also probably be higher than that which would be calculated on the basis of compensatory payments awarded to pollutees by courts, since they usually tend to give compensation for pain and suffering judged too small by the pollutees.

Finally, the damage cost may change with time as the polluted parties realise that the pollution is more or less serious than they had thought.

The law of the majority

The law of the majority is one way of reaching a decision when individuals have different opinions. If views on pollution amongst pollutees range from indifference to phobia, the damage cost for the polluted parties as a whole might be the excessively high if a single individual takes a too serious view of the damage. In practice, the damage cost can be estimated by determining the compensation necessary to win over a sufficient majority of the pollutees. If this majority authorises the establishment of the polluting enterprise, the international problem is reduced to a local one between the majority and the minority in the polluted area. Social tension is likely to run less high if the compensation paid is a public good (e.g., public park, zoo, theatre, centre for cultural activities, swimming pool) or if it benefits all the polluted parties equally (less local taxes) but could be higher if the majority retains the compensation for itself (by using it for its own purposes or by obtaining compensation in kind from which it is the principal beneficiary).

With the law of the majority, it is possible to make an estimate of the damage cost without it being necessary to win over all the polluted parties. In particular, pollutees may be compensated by a payment based on a formula adopted by the majority (for instance, 150 per cent of current values). This method avoids individual estimates being considered excessive by the community, but does involve some power of persuasion since it is a general consensus rather than the agreement of all the pollutees which is sought.

Contribution of information, persuasion and propaganda

Since the polluting country is liable to a tax which increases with the damage cost, its immediate concern is to try to minimise the damage cost perceived by the polluted country. In these circumstances it is not surprising that the polluters are accused, rightly or wrongly, of bias in their description of pollution phenomena. Community action groups in the polluted country will provide them with often excessive counter arguments. "Experts" will give conflicting views. In this context,

167

the polluted parties will be tempted to come out unanimously against any new polluting plant where they do not receive compensation, but will probably adopt a more balanced attitude if they are indemnified under the mutual compensation principle. It will then be in the interest of the authorities in the polluted country to determine the real (1) value of the damage (as perceived by the public) and to act as a moderating element between the polluters and their opponents. Since it is mainly a question of public acceptance, these authorities will in general try actually to consult the polluted parties and have them take part in decision-making. They will also avoid giving the impression that they are being influenced by any of the pressure groups.

Rejection of pollution

When the polluted parties adopt a highly negative attitude to pollution, they will be likely to declare a very high damage cost and the optimum pollution level will appear as if it were the reference level p_0 (and even sometimes lower than p_0).

If the polluting country accepts the judgment of the polluted country with regard to the extent of the damage, it will sometimes either have to bear high treatment costs or give up establishing a polluting enterprise near the frontier (2). This attitude on the part of the polluting country is tantamount to recognising the polluted country's right to accept or reject pollution above p_0, and maximises overall welfare.

If the polluting country does not accept the polluted country's judgment and gives permission for increased pollution without compensation, it creates an international problem and increases its own welfare while reducing that of the neighbouring country. This attitude on the part of the polluting country goes against the idea of maximising the common welfare since, so long as the polluted country is benevolent, the damage cost declared actually does correspond to reality. In fact, the polluting country rejects the game rules corresponding the co-operative and altruistic behaviour and seeks to maximise only its own welfare once it considers such game rules to be too binding.

Overbidding

The polluting country might sometimes suspect the polluted country of giving a high estimate of the damage cost in order:

a) to get higher compensation;

b) to benefit from special advantages to the polluting countries of the region in question due to specific geographical conditions;

c) to harm the polluting country.

The first hypothesis is clearly false since the polluted country benefits most by estimating the damage costs as accurately as possible (the mutual compensation principle is the only system having this property).

The second hypothesis corresponds to the wish of the polluted country to share - with no particular justification - a location rent (e. g., where the upper part of the lake of a hydroelectric project straddles a frontier) although being fully compensated. This presentation of the situation by the polluting country is inaccurate because, when accepting the level p_0, the polluting country recognized to the polluted country a right to receive compensation for damage and a part of the gain (surplus) resulting from operation at optimal level. As a matter of fact, the proposed argument has a role in the choice of p_0 which may be made very small if the polluting country draws an important benefit from the chosen location. Such negotiation on p_0 is outside the framework of the mutual compensation principle.

---

1) The situation is different when the polluted parties are compensated in relation to the vigour of their opposition. In this case, the authorities in the polluted countries will on occasions find themselves obliged to espouse the viewpoint of the polluted parties rather than looking for the best all round solution.

2) A better siting of the polluting enterprise may prove to be necessary and should not be viewed as unacceptable intervention, since environmentalists at national level are free to express their views and ensure that they prevail.

The third hypothesis is not covered in the above study (1). However, if a country were acting in a malevolent manner, it might well find it advantageous to give a too high estimate of the damage cost. When the polluting country is informed of the damage estimate made by the polluted country it might consider it as excessive if it believes the polluted country to be acting malevolently, but must consider it to be correct if it believes the polluted country to be benevolent. Since the polluting country is often in no position to detect malevolence, it sometimes has difficulty in assessing the correct assumption (benevolence or malevolence).

The polluting country will tend to believe that the polluted country is perhaps acting malevolently if it gives a damage estimate much greater than the one used by the polluting country within its own territory. Moreover, the polluting country might consider that it should not have to bear a polluting tax much higher than the one it would use for a similar case of national pollution and that it should not adopt a more favourable attitude to foreigners than to its own nationals. These arguments are not entirely valid if the foreign victims in no way profit from the polluting activity and if these foreign victims have a much higher social preference for the environment. Thus the polluted country runs the risk of finding itself accused of malevolence if it makes excessive demands, and hence runs the risk of seeing any agreement based on the mutual compensation principle being denounced. It could be induced to share the pollution control cost if it seeks a very low level of pollution (transfer from the polluted country to the polluting country owing to difference in social preferences) or if it wishes to avoid being accused of hampering the economic development of the areas along the frontier with the neighbouring country for reasons of undue environmental conservation. The polluted country will try to reject this solution it considers unjust, believing that it has the right to keep the environment in its initial state. The polluting country might retort that such a right is not recognised and in any case would be unacceptably harmful to the polluting country. Thus the compromise solution might finally be the lesser evil for countries having very different social preferences and might provide a way of avoiding the policy of the "fait accompli". If it appears that a financial compromise would be delicate to bring about for a particular case, the countries might negotiate the use of land in the whole frontier region as part of a long-term plan. They will then find the needed elements to reach an equilibrated solution without introducing transfers.

The reference level p_o

The reference level p_o is a reasonably low level of pollution, the choice of which is used to determine the initial situation and to avoid the need to analyse costs in conditions which, technically speaking, are not very well known. It could be a normal pollution level before a polluting undertaking is set up, or a normal pollution level in similar regions of the polluted country. This reference level is not an objective but a starting point for negotiations. Its choice may give rise to serious difficulties because of the lack of information regarding the costs and economic constraints it may possibly impose on the polluting country.

If the polluting enterprise does not have to pay the residual damage cost in the national context, this residual damage, being borne by the community, then the polluting enterprise near to a frontier would be governed by the same system as other polluting enterprises when the polluting country pays the residual damage cost and the enterprise a pollution tax with exemption. This direct contribution by the polluting country is justified since this country benefits from the fact that the damage is caused abroad. Contribution by the polluting government is not necessary when the polluting enterprise is liable to a pollution tax without exemption. However, the polluting enterprise may have to call for assistance from the polluting government when the cost of the damage caused abroad exceeds the damage cost for national pollution (installation grants in frontier regions).

The treatment cost

The problem of evaluating the treatment cost will not be gone into here although this is certainly a problem. Indeed the desire to site a polluting enterprise

---

1) Annex 4 gives an extension of the mutual compensation principle to malevolent countries. A country is said to be benevolent if its welfare does not increase when the welfare of a neighbouring country decreases and is said to be malevolent in the opposite case.

close to the frontier can be justified for many reasons which are difficult to quantify (balanced regional development, local unemployment, and so on) and it is rare that the frontier site is the only one which is technically feasible.

If the treatment cost were known by both countries, it would not be necessary to invoke the mutual compensation principle and it would be possible to use the treatment charge based upon the modified victim-pays principle (see report at pages 87-114).

Conclusion

Introduction of the mutual compensation principle makes a substantial positive contribution to the study of the case in question, and it would seem that it cannot be considered as worse than any other system. If the countries are benevolent, the proposed solution appears very interesting. However, to avoid compromising international relations and countries accusing each other of malevolence, the polluted country might well be induced to adopt a more flexible position when its social preferences for the environment are much higher than those of the polluting country. In this case, the polluted country might have to bear costs if it wished to keep its environment unchanged. Finally, the crux of the problem of the polluted country is to get its estimate of damage cost accepted while avoiding any accusation of malevolence and for the polluting country to get its estimate of treatment cost accepted while avoiding any accusation of malevolence. Thus it is a necessity to collect cooperatively quantified information on costs.

GENERALISATION AND EXTENSION

The purpose of this Annex is to show briefly that the principles which are described in this paper in relation to a particularly simple case can be generalised to more complex situations and to malevolent countries.

EXTENSION OF THE MODEL

Damage in the polluting country

When part of the damage is caused in the polluting country, the analysis can be generalised very simply by taking $C(p)$ to be the sum of the treatment and damage costs in the polluting country.

Treatment in the polluted country

When the pollution abatement measures have to be carried out in the polluted country, the problem of international efficiency does not arise since the polluted country will itself choose the optimal treatment level. The polluting country will be invited to bear a fraction of the total cost for reasons of international solidarity.

Treatment in both countries

If treatment has to be carried out in both countries, and if the damage is caused in both countries, the general model involves the following costs:

$C_1(p_o, p_1)$: cost of treatment in the upstream country to reduce pollution from level p_o to level p_1.

$D_1(p_1)$: damage cost when pollution is at the level p_1 in the upstream country.

$C_2(p_2 + bp_1, p_3)$: treatment cost in the downstream country to reduce pollution from the level $(bp_1 + p_2)$ to the level p_3 where p_2 is the pollution level caused by polluting activities in the downstream country if $p_1 = 0$.

$D_2(p_3)$: damage cost when pollution is at the level p_3 in the downstream country.

The mutual compensation principle invokes the following rules:

a) The joint agency looks for the values p_{1a}, p_{3a}, which minimises the sum:

$$\alpha_1 C_1(p_o, p_1) + \delta_1 D_1(p_1) + \alpha_2 C_2(p_2 + bp_1, p_3) + \delta_2 D_2(p_3)$$

where α_1, α_2, δ_1, δ_2, are the inaccuracy factors introduced by the countries.

b) The upstream country looks for the values of α_1 and δ_1 which minimise the net cost (direct cost plus tax):

$$T_M = C_1(p_o, p_{1a}) + D_1(p_{1a}) + \delta_2 C_2(p_2 + bp_{1a}, p_{3a}) + \delta_2 D_2(p_{3a}) - E_1$$

where p_{1a} and p_{3a} are functions of α_1, α_2, δ_1, δ_2, and E_1 is a constant or a function of α_2, δ_2.

c) The downstream country looks for the values of α_2, δ_2 which minimise the net cost (direct cost plus tax):

$$T_V = C_2(p_2 + bp_{1a}, p_{3a}) + D_2(p_{3a}) + \alpha_1 C_1(p_0, p_{1a}) + \delta_1 D_1(p_{1a}) - E_2$$

where E_2 is a constant or a function of α_1, δ_1.

When the calculations are made, it is found that the countries give the exact values of the costs ($\alpha_1 = \alpha_2 = \delta_1 = \delta_2 = 1$) and the treatment levels p_{1a}, p_{3a} are the optimal levels $p_1^{\textstyle *}$, $p_3^{\textstyle *}$.

The mutual compensation principle can thus be generalised to a much more comprehensive model than the one examined. For this it is sufficient to tax each country to take account of all the costs it does not bear in the first place, and to make the joint agency responsible for choosing the optimal pollution levels on the basis of information provided by the countries.

Mutual pollution

If two countries discharge amounts of effluent q_1 and q_2 into a common resource and can treat these effluents at costs $C_1(q_1)$ and $C_2(q_2)$, the total cost due to the pollution of the resource is the sum of the treatment and damage costs for the two countries:

$$C_1(q_1) + D_1(q_1, q_2) + C_2(q_2) + D_2(q_1, q_2)$$

If the mutual compensation principle is invoked, the joint agency is given information α_1 C_1, α_2 C_2, δ_1 D_1, δ_2 D_2 and selects the levels q_{1a}, q_{2a}, which minimise the sum of the costs declared. If each country is taxed according to the costs declared by the other country, it can be shown once again that it does not benefit the first country to declare an inexact figure for the costs it bears in the first place.($\alpha_1 = \alpha_2 = \delta_1 = \delta_2 = 1$). In these circumstances, the joint agency will choose the optimal discharge levels.

NUMBER OF COUNTRIES

The mutual compensation principle can be easily generalised to the case of three or four countries polluting a given river or lake. In theory, the principle can be used for any number of countries (for instance, air pollution in Europe), but in practice it could well be difficult to persuade a country to accept a tax which depends on costs in the countries with most of which it has no common frontier. When transfrontier pollution involves a great number of polluting and polluted countries, the problem can be analyzed as a situation of near perfect competition (large number of buyers and sellers of antipollution service).

EXTENSION TO THE CASE OF MALEVOLENT COUNTRIES

The analysis given in this paper concerns two countries which are trying to minimise the net cost of transfrontier pollution that they each bear:

$$\underset{\alpha}{\text{Min}} \left[T_M(\alpha, \delta) \right]$$

$$\underset{\delta}{\text{Min}} \left[T_V(\alpha, \delta) \right]$$

on the assumption that the countries are benevolent.

If the countries feel better off when the neighbouring country suffers a reduction in wealth (envy, jealousy, hostility and so on), the assumption of benevolence is no longer justified and the theory set out above does not apply.

When the two countries are equally malevolent, it may be assumed that each is trying to minimise the sum of its own net cost and the additional net cost it bears compared with the net cost of the other country:

Upstream country: $\quad$ Min $\left[T_M + n\,(T_M - T_V)\right]$

$\qquad\qquad\qquad\qquad\qquad\qquad\qquad\qquad n > 0$

Downstream country: $\quad$ Min $\left[T_V + n\,(T_V - T_M)\right]$

where n is the coefficient of malevolence (n = 0 means that the countries are benevolent).

When the objective is defined in this way, the mutual compensation principle can be generalised by applying a tax $\left[n\,\alpha\,C + (1 + n)\,\delta\,D - E_1\right]$ to the polluting country and a tax $\left[n\,\delta\,D + (1 + n)\,\alpha C - E_2\right]$ to the polluted country so that the net costs are

$$
\begin{cases}
T_M = C + n\,\alpha\,C + (1 + n)\,\delta\,D - E_1 \\[2mm]
T_V = D + n\,\delta\,D + (1 + n)\,\alpha C - E_2
\end{cases}
$$

where E_1 and E_2 are two lump sum transfers independent of p, α and δ.

The joint agency is informed of the costs αC and δD and chooses the pollution level p which minimises $(\alpha C + \delta D)$. The polluting country estimates that the cost T_V is $(1 + n)\,(\delta D + \alpha C)$ since a priori it does not know the real damage cost. Similarly, the polluted country estimates that the cost T_M is $(1 + n)\,(\alpha C + \delta D)$. Here the objective function of the polluting country is $(1 + n)\,(C + \delta\,D)$ which passes through a minimum for $\alpha = 1$. Similarly, the objective of the polluted country has a minimum for $\delta = 1$. Finally, the two countries will reveal the exact costs and the pollution level chosen by the joint agency will be the optimal level.

When the two countries are malevolent, but to different degrees, it may be assumed that they minimise the following functions:

Upstream country: $\quad$ Min $\left[T_M + \dfrac{m}{1 + n - m}\,(T_M - T_V)\right]$

Downstream country: $\quad$ Min $\left[T_V + \dfrac{n}{1 + m - n}\,(T_V - T_M)\right]$

where $0 \leqslant /m - n/ < 1$

If the taxes are

$\left[m\,\alpha\,C + (1 + m)\,\delta\,D - E_1\right]$ and $\left[n\delta\,D + (1 + n)\,\alpha\,C - E_2\right]$,

the net costs are:

$$
\begin{cases}
T_M = C + m\,\alpha C + (1 + m)\delta\,D - E_1 \\[2mm]
T_V = D + n\delta\,D + (1 + n)\alpha C - E_2
\end{cases}
$$

It can be shown, as above that the polluting country minimises the cost $(C + \delta\,D)$ and the polluted country the cost $(D + \alpha\,C)$. They will choose the values $\alpha = 1$ and $\delta = 1$, and the pollution level chosen by the joint agency will be the optimal level, i.e., the level which minimises the total cost $(C + D)$.

Remarks

a) The degrees of malevolence are the data upon which the countries will find great difficulty in agreeing. Using zero values (assumption of benevolence) leads to suboptimal pollution levels if the countries are in fact malevolent.

b) It does not seem possible to set up a system to guarantee economic efficiency irrespectively of the values of the coefficients of malevolence.

Let the utilities of the upstream and downstream countries be

$$U_M \left[A(p_o) - C(p, p_o)\right]$$

$$U_V \left[B(p_o) - D(p, p_o)\right]$$

where $C(p_o, p_o) = 0$ and $D(p_o, p_o) = 0$

$$\frac{\partial C}{\partial p} < 0 \quad \text{and} \quad \frac{\partial D}{\partial p} > 0$$

At the level p_o, the utilities are $U_M \left[A(p_o)\right]$ and $U_V \left[B(p_o)\right]$. Let us assume that there is a pollution level p^* which maximises the expression

$$\beta_M U_M(p, p_o) + \beta_V U_V (p, po)$$

which reflects the joint objective.

Expand U_M and U_V as a series:

$$U_M \left[A(p_o) - C(p, p_o)\right] = U_M \left[A(po)\right] - \left.\frac{\partial U_M(X)}{\partial X}\right|_{X = A(p_o)} C(p, p_o) + \ldots$$

$$U_V \left[B(p_o) - D(p, p_o)\right] = U_V \left[B(po)\right] - \left.\frac{\partial U_V(X)}{\partial X}\right|_{X = B(p_o)} D(p, p_o) + \ldots$$

Let $\frac{\partial U_M}{\partial X}$ and $\frac{\partial U_V}{\partial X}$ be constant in the interval $p^* - p_o$ (marginal utilities constant).

The sum of the utilities is then maximised by minimising the sum

$$\beta_M \left.\frac{\partial U_M}{\partial X}\right|_{X=A(p_o)} C(p_1 \, p_o) + \beta_V \left.\frac{\partial U_V}{\partial X}\right|_{X=B(p_o)} D(p, p_o)$$

Let $x = \dfrac{\beta_V \dfrac{\partial U_V}{\partial X}}{\beta_M \dfrac{\partial U_M}{\partial X}}$ and let taxes $x \, \delta \, D(p, p_o)$ and $\frac{\alpha}{x} C(p, p_o)$ be imposed on the upstream and downstream countries respectively. The pollution level p_a is chosen by the joint agency to minimise the sum

$$\alpha \, C(p, p_o) + x \, \delta \, D(p, p_o)$$

The upstream country chooses the value of α which maximises

$$U_M \left[A(p_o) - C(p_a, p_o) - x \, \delta \, D(p_a, p_o)\right]$$

Since $\quad -\frac{\partial U_M}{\partial \alpha} = \frac{\partial U_M(X)}{\partial X} \left[\frac{\partial C(p_a, p_o)}{\partial P_a} + x \, \delta \, \frac{\partial D(p_a, p_o)}{\partial P_a}\right] \frac{\partial Pa}{\partial \alpha}$

the upstream country chooses $\alpha = 1$. Similarly the downstream country chooses $\delta = 1$. Under these conditions, the level p_a minimises the sum of the costs

$$C(p, p_o) + x \, D(p, p_o)$$

and maximises the joint objective. The level p_o is therefore the optimum level p^*. It should be noted that the joint agency does not minimise the sum total of the taxes it receives and that the agency has to know the quantity x, i.e. the ratio of the marginal utilities of the two countries. Thus it appears that agreement is possible only if the countries agree on the value of marginal utilities and on the weights β_m and β_V. Otherwise, they could choose $x = 1$ (as in the rest of the report), but this would be unfair when the levels of development are very different. We may note that x depends on p_0, i.e. on the equity principle adopted, but that the mutual compensation principle is still applicable.

Part Three

INSTITUTIONS

TRANSFRONTIER POLLUTION: ARE NEW INSTITUTIONS NECESSARY? [*]

by

A. Scott

Prof. of Economics, University of British Columbia, Vancouver
Canada

[*] All parts of this paper have been written with the assistance, and close
collaboration, of Peter Gardner.

177

SUMMARY AND INTRODUCTION

Summary

Transfrontier pollution (TFP) is an externality. Economists give it this name because it gives rise to external costs - costs that do not enter into the choices made by the originating country about how much TFP to permit. Because the campaign against TFP is new to international affairs, textbook examples of economic institutions that can channel the effort at minimum cost are rarely encountered. This is not to say that there are no examples of accepted international law or even organisations dedicated to making the best of externalities. Maritime law as evidenced by the forthcoming Law of the Sea Conference and the existence of IMCO; air law, as evidenced by the Warsaw Convention and the existence of ICAO; radio-spectrum law, as evidenced by the Moscow Conference and the ITU; and outer-space law all offer precedents. The control of contagious disease through WHO is even closer to our subject.

Large-scale physical externalities between adjoining countries, however, have been dealt with comparatively rarely. One reason may be that such examples of types of spillovers are not numerous. Fisheries and navigation rights are the oldest example, and they have long been built into peace treaties and trade pacts. Until recently they rarely called for fisheries management however; their main intent was to enable fishing grounds to be treated among nations as essentially national territory, reserved for special economic groups (1). The best spillover example, indeed, is in rivers used by two or more states, for the navigation, consumption, levels and flows of which a body of treaty law has slowly emerged (2). But pollution of such rivers has not been considered to be a serious problem, and both general international law and treaty provisions have been slow to adapt to this issue.

This paper is intended to deal with part of this problem, borrowing much from a number of papers sponsored by OECD. It first develops a model within which the choice of types of international institutional organisation can be placed. It then surveys the institutional choice itself. This subject is no longer restricted to OECD. Attention should be drawn to recent papers by D'Arge and Kneese, Blackmore, and others who are not presenting their thoughts to the OECD group.

Conclusions and summaries are inserted at a number of places. Among the most interesting are the following:

1. Although it is tempting for many economists to talk about developing agencies within which TFP decisions can be "internalised", the essential role of such organisations should be seen as the production of technical, economic and social data which is accepted by both countries without question.

2. Transfers, payments or payment-in-kind will almost certainly be required in most agreements.

3. Formal bargaining about agreed levels of TFP, and payments, is unavoidable, even if this leads to unpleasant argument about which country

---

1) See F. T. Christy Jr., and Anthony Scott, The Common Wealth of Ocean Fisheries, Baltimore, Johns Hopkins Press, 1965.

2) See Chapman, J. D., The International River Basin - Proceedings of a Seminar on the Development and Administration of the International River Basin, University of British Columbia, Publications Centre, 1963; Bourne, C. B., "The Right to Utilize the Waters of International Rivers", Canadian Yearbook of International Law, 1965, pp. 187-264, and later papers by the same author; and Smith, H. A., The Economic Use of International Rivers, London, P. S. King and Son Ltd., 1931.

has "rights" to the river. Such devices as cost-sharing (CSA) (1) which it has been hoped might prevent such unpleasantness from arising, present such problems of their own as to be unworkable.

4. The theoretical model deals not only with upstream-downstream pollution problems, but also with reciprocal "lake" problems. This model suggests that the chief role of international cooperation on a lake is not to bring about a state of the lake that would never be achieved by independent action, but to do so more quickly, with less uncertainty, and with greater equity.

5. For each country, the best institution is one that minimises the total costs of TFP. This can also be rendered as one that maximises the benefits of TFP abatement. The costs referred to include Abatement Costs (AC), Damage Costs (DC), transaction costs between jurisdictions and agencies, agreement costs within jurisdictions and agencies and compliance costs between such jurisdictions and private actors.

6. From the point of view of two or more countries combined, the best set of institutions and policies is one that minimises the international total of such costs. Their distribution must also be considered, and indeed may be the main preoccupation of most experts other than economists and natural scientists. (For the world as a whole, "damages" include the damages that flow from the basin occupied by the two or more nations to the global ocean or atmosphere. Such "global" effects, which weigh heavily in the Stockholm discussions, are not discussed in this paper.)

7. From the international point of view, the best institutional arrangement is not one that minimises AC plus DC. Nor is it one that minimises administrative, diplomatic and political troubles and worries (that is transaction costs, agreement costs and compliance costs). The best agreement brings about a set of institutions that minimises the sum of all these costs.

o

o   o

Ideas about the setting up of institutions to deal with Transfrontier Pollution (TFP) will come from various sources. International Law and Relations is one; engineering is another; and public health and sanitation a third. The Stockholm conference showed too that conservationists and environmentalists per se have important suggestions to contribute. This paper is mostly about the contributions that stem from economics.

As with any social science, economics often confuses people by its failure to disentangle the various elements that make up its approach. These may be worth listing separately at this early stage:

1. The fundamental ideas of economics that do not turn up elsewhere: marginalism, the market, opportunity costs, market failure.

2. Predictive statements about what is, or is likely to happen (positivism) (see pages 181-186).

3. The recommendation of policies based on combined aims about what ought to be done (idealism or normative economics) (see pages 187 and 191).

The economic scrutiny of institutions for dealing with TFP requires the application of all three of these elements. Marginalism, etc., is inherent not only in the emphasis now being given to environmental objectives but also in the other aims of the nations: justice and fairness, national sovereignty, economic efficiency in the use of labour and capital; growth; stability; and so forth; while positivism (and objectivity) is necessary in recommending the various alternative sets of procedures and institutions that might be used to achieve as much as possible of the aims that society sets for itself.

---

1) See report by Muraro in "Problems in transfrontier pollution", OECD, 1974.

The Fundamental Contributions of Economics

In the present context, economics has four fundamental ideas that are worth bringing to bear, the first of which is marginalism. This idea stems from the economists' experience that it is rarely desirable to proceed as far as physically or socially possible in any direction. Just as with lack of education, disease, or crime, so with pollution: there is some best degree of application of policies that would eradicate these conditions. To go further is to run into rising costs or sacrifices of other things, while perhaps getting little additional benefit. Thus, while environmentalism and ecology appear to urge mankind to stamp out pollution totally, economics counsels the search for that amount ("margin") beyond which further pollution-abatement activity would cost more than it would be worth.

The implication for the present inquiry is that we are not looking simply for institutions by which neighbouring nations can cooperate to clean up their shared environment, but also for the institutions by which they can learn, and agree on, the right amount of clean-up. We envisage them as searching for that margin (level, amount) beyond which it would not be worth-while for them to go, taking everything into account.

The second fundamental idea of economics is that of "opportunity cost". It is that the drawback to proceeding with anti-pollution policies (or any other aims) is not simply in the apparent cash expenses of doing so. It is also in the non-monetary burdens, costs and sacrifices that will be entailed. These include "intangible" costs or losses that are not measured in the market place, of socially valuable recreation, modes of living, open spaces, and so forth. More fundamentally, it is that all costs, whether or not measured in the market, arise basically from opportunities foregone or sacrificed. For example, the disappearance of certain "dirty" techniques or products, and the installation of pollution-abating devices, ultimately requires that consumers get less of certain products and leisure than they now enjoy, as labour and capital are re-allocated from some of their present employments. These losses may show up in higher prices of things made more scarce, if such things are sold through a market; if not, they will simply be felt as losses, sacrifices, or shortages suffered by some people. All of them, marketed or not, are the "real" or "opportunity" costs of environmental policies. The implication for the present inquiry, of course, is that nations do (and should) ascertain the real or opportunity costs of joint environmental action, in discovering how far to proceed, not simply the apparent cash expenses. As everyone knows, these may not coincide.

The third of these fundamental ideas is that of "market failure". The economist tends to think that individualistic behaviour, channeled in a competition market place, will lead to an efficient allocation of labour, capital and natural resources among goods and services.

However, the market is unequal to this allocational task in many circumstances, including most of those involving the quality of the environment. Among these are:

i) "spillovers" or "externalities". Here benefits are conferred or costs are imposed outside a given economic unit, without relief or compensation. These can arise when the effects are felt within a "common property" resource, or when a resource that might well be aggregated into independent, private units is divided by a national frontier.

ii) "Public goods". Individuals or firms will fail to produce certain goods and services if the beneficiaries can be "free riders" - need not pay for the services received. For example, only government will find it worthwhile to provide environmental clean-up services that are equally accessible to all consumers, because no private charge can be successfully levied, exclusion being feasible.

Thus, to summarize, the distinctive aspects of an economic approach to the study of institutions for controlling and paying for TFP can be contrasted with the approaches that would be adopted by other specialists.

i) In contrast with an all-or-nothing approach to pollution, the economists' instinct is to search for some institution that will lead to a just balance between the costs of abatement and the costs of degradation.

180

ii) In contrast with the approach of the accountant or engineer, the economists' instinct is to look to alternatives. The costs of abatement and pollution are therefore to be measured not by actual outlays but by what might been done instead, by alternatives, or by the value of what has been sacrificed or foregone.

iii) In contrast with the approach of many planners and administrators, the economist tends to assume that in the absence of regulations and controls, people and firms will act independently in a way that is advantageous to everyone. However, when certain causes of "market failure" are present, a prima facie reason for intervention exists.

Thus we seek to repair the harm done by spillovers and the lack of marketability of pollution abatement services, given that the action's advantages (reduction in real costs) should exceed its real burdens, and given that the best action is likely to be an institution or procedure that leads to some compromise between the extremes of high pollution with heavy damage costs and high purity with heavy abatement costs.

A MODEL OF T.F.P. POLLUTION

The purpose of this Part is to use both positive and normative economic reasoning to analyse TFP situations (listed as points 2 and 3 in Part One). First we wish to predict the outcomes of independent and combined action by neighbouring countries. Then we shall compare these outcomes with the normative concepts of the "ideal" abatement of TFP.

Upstream-Downstream TFP

We begin with discussion of the economic relation of two countries on an international river. We postpone till Section 3 the reciprocal TFP of a shared lake or sea.

The Prediction of Independent Action and Reaction

Positive predictions about national actions may divide our predictions into two categories: what is likely to happen in the absence of international agreement on TFP; and what is likely to happen during the process of agreeing and follow-through. It is obvious that the second is dependent on the first.

In the absence of specific agreement on TFP, any country's actions can be predicted to depend on its rights. Most nations' laws and actions reflect a belief that they have an unlimited right to use any water or air from which they are not excluded for waste disposal. It is possible, using international law doctrines about the duty of a country to take into account the effect of its actions on other countries, to argue that the waste-disposal right is not so firm as is believed. That is, it can be argued that downstream, or victim, countries, also have rights. But (apart from bilateral treaty situations in which downstream rights are specified) a right to unrestrained waste disposal appears to have been assumed.

On this assumption about rights, it follows that a country will dispose of wastes into water (international or domestic) as long as it pays to do so. The limits to its waste discharge into water will be determined by the lowest cost of some alternative method. These include:

i) the alternative costs of re-cycling wastes;

ii) the alternative costs of emitting wastes into other water resources, the air or the earth;

iii) the loss of profit from foregoing production altogether (where "profit" must be interpreted to include corporate net profits, direct taxes and extra consumer surplus);

iv) and the loss of profit from moving production to another location.

The least of these four types of cost is the relevant alternative national social abatement cost (control cost), AC, of reducing waste discharge.

To these four alternative abatement costs must be added the local damage cost, DC, of waste discharge (1). Because a country will not transmit to its neighbours more water pollution than it can stand itself, we may say that the limits to its waste discharges will be set by the total costs that pollution imposes on it: the sum of abatement costs and damage costs. To summarise: in the absence of transitional changes, the steady state condition of a river or lake will be that set by a nation that is reducing to a minimum the total costs, TC, of pollution, including in the term "costs" both the expenses of dealing with waste in some other way and the damage from not enjoying a cleaner environment.

The discussion so far, about rights, abatement costs and damage costs applies directly to the prediction of the amount of pollution that a nation will permit in a domestic lake or river, wherever the water eventually flows. It explains therefore, the behaviour of an upstream nation, hereafter called A; such a nation will be influenced by the total of its domestic abatement and damage costs, and uninfluenced by the downstream effect of its discharges. It will continue to emit discharges that minimise its total pollution costs (AC + DC) leading to a given transfrontier pollution q_A^i.

The typical behaviour of a downstream country, however, does require analysis. First we may distinguish two categories. In the first, a single downstream country must behave completely passively. It cannot affect the TFP. In the second category, there are two or more downstream countries. We deal with this category in a later section.

In the first category, the downstream nation, B, will receive the pollution from upstream, reduced by whatever dilution, degradation or assimilation takes place in the course of transmission between the two regions. This adds to the pollution created by its own discharges. The downstream abatement cost (AC) of a given amount of pollution will therefore be much higher than if it did not receive TFP from upstream, since the only techniques available for direct TFP abatement will require the mass downstream treatment (or dilution) of the river itself. However, the total level of the augmented downstream pollution can of course be reduced by abating more domestic pollution. The local damage cost (DC) of a given amount of total pollution will be the same as it would have been without TFP. To minimise the sum of these two costs will probably require the downstream state to

---

1) "Damage cost" is a term of art, or jargon, that is widely used but irritatingly misleading. In most analyses the aim is to find the total amount that victims of (or sufferers from) water pollution would pay to be free of the pollution, at a given time and place. This "willingness to pay" is then assumed to stem from the victim's valuation of the "damage" or harm that the unabated pollution imposes on him. Of course, willingness to pay may be much less than damage cost, if some lower expenditure would purify the water a victim uses or in other ways relieve him of the "damage"; or be much higher if society's collective distaste for environmental or ecological disruption exceeds the sum of individual estimates of pollution costs. Some authors avoid the "damage cost" terminology by referring instead to the value of "benefits" or pollution abatement. And some writers simply confine themselves to the notions of "demand" for abatement, or "willingness to pay", leaving unspecified the causes or motives for these amounts.
A later section of this study will review practical difficulties of ascertaining actual measurement costs and will recommend that their amount not play an important part in any international TFP procedure and, in fact, the procedure should not be made to depend on ascertaining their amount.
But any economic discussion must be based on some general belief about whether additional pollution evokes increasing, constant, or diminishing willingness to pay. Consequently, at this stage we assume that a "damage cost" function exists, that it is measurable, and that pollution clean-up elicits diminishing willingness to pay.

suffer a higher aggregate amount of pollution than it would tolerate if all discharges could be treated at the source within its own jurisdiction (1).

As we will see, the burden of total cost TFP will motivate the downstream state's desire to obtain rights that will protect it from TFP, or to obtain an agreement to mitigate its total damage and abatement costs. While this burden cannot be directly observed (because there is no way of knowing about its public and private actions in the absence of TFP), a few commonly made assumptions will enable us to make three predictions about its reaction (adjustment) to TFP:

    i) TFP will not cause it to give up local abatement altogether; probably it will abate <u>more</u> of its own pollution than in the absence of TFP;

---

1) This discussion may seem curiously indirect to lawyers. To them it will seem obvious that the downstream country suffers from excessive pollution simply because it has no jurisdiction over the upstream sources. They are right, of course. The treatment in the text above simply takes account of the fact that a downstream country actually has two choices:

    i) whether to abate pollution originating within its own frontiers,
    ii) whether to invest in abating the TFP that it receives.

It is not obvious that legal jurisdiction alone would cause it to do either. Low costs may motivate it to do both; high costs may prevent any action. Hence, given the rights of the upstream country, costs explain the eventual determination of the amount of downstream pollution. Equally, given the costs of abatement at various points between source and downstream victim, the extent of the downstream nation's jurisdiction determines the amount of pollution that it will suffer.

Figure 1 shows the upstream and downstream positions. The pollution q_A^i, determined upstream, becomes the TFP of the downstream state. Downstream pollution in the absence of TFP, would have been the amount shown at least cost point C_B. TFP raises it to $q_A^i + q_B^i$. As TFP rises, B's downstream least cost point rises from C to C', to C''.

The 'commonly made assumptions' are that marginal abatement costs are decreasing and that marginal damage costs are increasing functions of the amount of pollution. The former assumption is reasonable because pollution abatement at some locations is surely different from that at others, for many reasons; consequently, if these costs are arrayed in order, reducing the amount of abatement requires the country to incur increasingly smaller average and marginal costs.

The assumption that marginal damage costs increase is trickier, and is wholly unverified for any important stream. It stems from the theory of diminishing utility. A person on a heavily polluted river is assumed to benefit more from a small improvement in water quality than such a person would from the same improvement in a nearly clean river. That is, benefits from pollution abatement is characterised by diminishing returns. A version of this is widely used in demand theory, where it may be expressed by saying that the more of a good a person is enjoying (e.g. pollution abatement), the less of any other good he is willing to give up to get a unit more of it. A similar theory may be applied to an industrial consumer.

The problem with the assumption is that its opposite is also attractive. A person on a heavily polluted river may be assumed to be injured to the misery of dirty water, and will pay little to obtain a small improvement; while the same person on a clean river may gain greatly from a small further improvement. If the damage curve is linear, both theories cannot be correct. The conclusions made in the text follow most definitely if there are diminishing marginal benefits from pollution abatement; they also follow if marginal benefits are constant. The two contradictory theories can be reconciled if we assume that the marginal damage curve is like an inverted U. At its highest point is that amount of pollution for which the person gets the greatest benefit, when quality is slightly improved. At either end, however, small changes in quality are much less important to him. This means that DC, with respect to pollution, is an ogive, or S-shaped. However, almost any upward-sloping DC curve will illustrate our main points.

ii) however, it will experience more total pollution (its own and TFP) than without TFP;

iii) consequently, its total costs - damage and abatement - will be higher than in the absence of TFP.

Prediction (ii) can be most conveniently summarized by reference to a reaction curve (1). The first panel of Figure 2 shows the upstream country's generation of pollution q_A^1. This comes as TFP to country B, and B's reaction is shown by the heavy line. The total amount of pollution in the river in region B ($q_A + q_B$) is shown by the numbered dashed contour lines.

Prediction (iii) is extremely important. These rising extra costs (C C'C'' in Figure 1 Panel 2) measure the total cost of pollution in B after all domestic adjustments to TFP ob abatement and consumption have been made. Their increase provides the incentive for B to avoid them by bargaining with A, or by taking other international action.

The Prediction of Bargaining Outcomes

We are now ready to use the apparatus so far developed to predict the outcome of bargaining between A and B. This second category of prediction concerns what is likely to happen during the course of agreeing. Because this is also the subject of ensuing sections, the discussion here can be comparatively brief.

Let us suppose that the nations decide that the disadvantages of allowing TFP to continue outweigh the advantages. They may be supposed to open discussions looking forward to coordinated or combined action. Then they will be observed to be searching for some formula that will make each of them better off with an agreement than it would be without it. This statement is tautological of course, merely implying that the upstream country will not take action unless failure to do so would make her worse off (in terms of her trade, defense, transportation and other arrangements with her neighbour, in terms of international law, or in view of some inducement the neighbouring country is offering); and that the downstream country will also not act unless her total outlays bring an advantage in reduction in TFP at least as valuable as other uses she might make of the same domestic resources and finances. Each will be uncertain at the outset about what information about damage and costs to stress until some general principles are laid down and accepted. Until these rules exist, neither knows whether to complain or keep silent about its various costs. For example: if they believe that reduced TFP and the polluter-pays principle will both prevail, a downstream country will tend to mention and even exaggerate the damage it is suffering, while an upstream country will tend to emphasize the high costs of abating pollution. But if a rule is to be adopted that requires the beneficiary to pay for its gains, the downstream country will have a smaller incentive to claim severe damages from TFP, and if there is no prior agreement on these principles, each will be reluctant to provide any information at all.

Each, of course, will have some notion about what sort of final agreement it is likely to obtain from its neighbour, as well as some notion about the general order of magnitude of both abatement costs and damages. In the light of these notions each will press for the adoption of a general rule which it feels will enable it to squeeze the greatest advantage from whatever data eventually are accepted.

In general, therefore, the downstream country can be expected to press for a system that (i) reduces its damage from TFP (allowing it both economic development and more enjoyable consumption of water quality); (ii) increases the compensation it obtains for past or future total or residual damage from TFP; (iii) reduces the inducement, if any, that it must pay to the upstream country; and (iv) minimises any implication that the upstream country has continued rights to dispose of waste at any particular level.

---

1) The reaction curve will be linear if B's marginal damage and marginal abatement cost curves are linear. If (in line with discussion in the previous footnotes) B's damage curve shows constant marginal benefits from changes in pollution, then its reaction curve will be vertical, perpendicular to the B axis.

   A country's marginal damage curve would have to be vertical to cause it to react by cutting its own emission by the same amount as an increase in TFP from the other country.

The upstream country, of course, will seek principles that work in the opposite direction. It will promote the adoption of a system that (i) permits it to continue using the stream for waste disposal; (ii) minimises the amounts, if any, it must pay as compensation for past or future total or residual damage; (iii) increases the contribution it receives from downstream to pay abatement costs; (iv) minimises any implication that any agreement reached, limits its future rights to dispose freely of wastes in the stream.

Of course, neither country can obtain all these aims, nor are they compatible. Each will bargain, trading off one aim against the other, until the general principles are adopted.

Once the principles are adopted, each can be expected to produce information (data, claims, accusations) which will tend to minimise its costs (maximise its gains) under these principles. This prediction cannot be over-emphasized. Neither will have an incentive to produce accurate estimates. It follows that one of the greatest gains from thoughtful planning and organisation of TFP institutions is the production and utilization of accurate information.

Finally, it is important to remember the importance to each nation of independence and self-determination. National freedom of action is perceived to be an end in itself, well worth sacrificing other net advantages for. Thus, as already mentioned, countries will be loath to enter agreements that imply a permanent concession of their rights (to discharge wastes, or to receive pollution-free waters) because this will be seen as a direct loss of national strength or power. Again, they will tend to depreciate systems of TFP abatement or control that require (on a continuing basis) that foreigners are visibly involved in monitoring their performance, measuring their water quality, or making decisions that affect their daily lives or industrial structure.

The general bargaining outcome can be predicted diagrammatically by using Figure 2. It will be shown that some idea of rights must exist, however, before bargaining can succeed.

The third panel of Figure 2 shows CC_B increasing as a function of q_A. Probably it increases at an increasing rate; this curvature is exaggerated in the figure. The CC_B curve cuts q_A^i at M; M is the total additional cost suffered by B because A is discharging pollution at A's least cost (optimum) point. Thus the declining vertical distance between CC_B and M is a measure of the decline of B's rent or surplus as TFP increases from zero to q_A^i. In the ensuing discussion, in panel 4, this declining rent is plotted as GG_B, both in the upper quadrant and the lower quadrant. Also plotted is that portion of TC_A lying below the q_A^i level of pollution. In the upper quadrant, by choice of origin, both are made to equal zero at q_A^i.

First, assume that A, the upstream country has full rights to discharge pollution. In the absence of bargaining, it will choose to be at least cost point q_A^i.

To see the possibilities of bargaining, we may look at the offers from B, for coordinates of points in the upper quadrant may be regarded as payments to the "landlord", A, to move back from q_A^i to some lower TFP. Since TC_A shows A's costs of making such changes, to be successful the offers must be depicted by points at least as great as TC_A. That is, the vertical distance from any point down to TC_A shows A's gain (loss) from abating pollution to that extent. Points on GG_B, showing B's total gain from abatement to that extent, represent the maximum offer B can make. Hence we can make the following deductions:

i) When A has rights, and is the landlord, B can afford to make attractive offers only for those ranges of pollution where GG_B is at least as great as TC_A. Such ranges may not exist; B's gains may everywhere fall below A's costs.

ii) The vertical distance between GG_B and TC_A indicates a bargaining range. If we draw parallel curves to GG_B and TC_A, indicating iso-revenue contours, we can indicate certain points: N, where A extracts a maximum all-or-nothing payment from B; and Q, where A's costs of abating pollution are just covered by B, analogous to the application of the victim-pays-principle.

Next, assume that B has the necessary rights to prevent the upstream country discharging into the river, or to charge for pollution. Then the same curves must be reinterpreted as changes in gain or loss from an initial position where

185

there is zero TFP. Thus, measuring downwards from the q_A axis, we see that the vertical relationship between the two curves has changed, they having been brought into equality at zero TFP. As TFP increases, the increasing vertical distance from the axis shows that A's gain increases (costs fall); while B's costs rise. (Actually, in the lower quadrant, the GG_B^1 curve may be regarded either as a downward shift of GG_B, or a plotting of CC_B in the negative direction.) The following deductions can be made:

    i) If GG_B^1 lies vertically below TC_A^1, A cannot profitably offer enough to induce B to permit any TFP.

    ii) The vertical distance between GG_B^1 and TC_A^1 indicates a bargaining range. If we draw iso-revenue curves parallel to GG_B^1 and TC_A^1, we note two points: P, where B extracts an all-or-nothing maximum revenue from A; and R, where B's losses from TFP are just covered by A, analogous to the application of the "Civil Liability" principle for that amount of pollution.

On certain assumptions about the curves TC_A and GG_B, the points N, Q, R and P may be vertically aligned. That is, regardless of which country holds the initial rights, and regardless of the strength of its bargaining power, the amount of pollution or abatement bought will be fixed. But such assumptions unnecessarily restrict the generality of the analysis. In general, the locus NQRP can slope toward or away from the vertical axis. (1)

"Cost sharing" as an alternative to bargaining

It has frequently been suggested in the OECD TFP papers that the bargaining-and-rights approach (see page 184 above) should be abandoned. Some of the reasons (which have to do with uncertainty about pollution rights in international law) will be outlined in a later section.

The suggested replacement for bargaining is some version of a "cost sharing" approach (CSA). This has been defined in various ways. In the report at pp. 31 it appears simply to be distinguished from some diplomatic or international law approach in which the countries do not deal or bargain with each other. Here however, it has a more definite meaning which makes it an alternative to the bargaining approach just outlined.

Instead of making and receiving offers leading to a change in the level of TFP (that is, buying and selling rights to TFP), the parties are envisaged in the CSA approach as making an initial agreement about their percentage shares of the total or pooled costs of both countries (these percentages can vary with various indicators; in the PPP case below, one country pays all the abatement costs, or some fixed share of them). While details differ among versions, the proposals usually contemplate that, with the shares settled, the parties should proceed to agree on the level of TFP, or abatement to be jointly provided and enjoyed.

The analysis of this approach employs Figure 3 (the same as Figure 2 used in the bargaining approach). The total costs, for various amounts of TFP, are measured vertically between GG_B^1 and TC_A. These are the combined costs of pollution that are to be shared.

    i) The polluter-pays principle (PPP). This most-frequently cited arrangement is not always recognized as a sharing principle. It embodies an idea popular in shared international projects: each side undertakes the works or adjustment necessary on its own side of the frontier. (When this is not thought equitable,

1) Much of this follows Muraro's excellent paper in Problems in TFP, OECD, 1974. One specific difference is in the derivation of the GG_B curve. Another is the treatment of GG_B as a net damage curve so that, on the assumption that the marginal utility of income is constant, curves parallel to it can be treated as indifference curves. In the text above they are treated as iso-revenue curves. In any case they stem from production functions as well as from utility curves.

If the damage curves were based wholly on utility functions with income and pollution as arguments, a constant marginal utility of income, equal in the two countries, would be a sufficient condition for a vertical "contract curve" like NQRP. (See report given pp. 31-54)

features may be added to the project on one side of the border until a more acceptable distribution of burdens is achieved.) In the present case, the shared costs locus runs along the q_A axis. The abatement etc. costs above the axis are assigned to A, while the damage etc. costs below the axis are borne by B.

The outcome of this initial principle is indeterminate. A is given an incentive to minimise its share of combined costs by not abating pollution at all, moving the level of TFP to q_A^1. B, on the other hand, is given an incentive to move to zero TFP. The PPP principle used in a CSA context, will not resolve the conflicting aims of the countries.

Of course, the conflict could be resolved by the two countries bargaining to some compromise level of TFP. If A had prevailed, B could "bribe" it to reduce its level of pollution. But, as soon as bargaining commences, the initial shares of the two countries are being altered. Bargaining, therefore, and not cost sharing, is the instrument of TFP resolution. The PPP cannot be successfully used as a prior cost sharing principle, as defined here.

ii) Sharing abatement costs. The upstream nation may agree to pay a fixed percentage of actual abatement costs, while the rest of combined abatement and damage costs are borne by the downstream party. This sharing principle, like the first, gives the upstream nation an incentive to advocate zero abatement. The downstream nation's incentive to advocate abatement will depend on the share of the abatement cost it must bear; the greater its share, the closer it will wish to move to the minimum combined cost, or "optimum". Neither of these first two cost sharing principles will assist the two countries to converge on any mutually acceptable level of TFP, "optimum" or otherwise. Their interests under these principles, diverge.

iii) Fixed share. One nation may agree to pay a fixed amount of the combined costs. On this basis, the second country would then seek that amount of abatement of TFP that would minimise the remaining combined costs. This outcome would be the same whatever the fixed amount to be contributed by the first country (for example, if it were small (or zero) the second country would in a sense internalise all the costs). It would choose the minimum combined cost ("optimum") TFP.

iv) Fixed percentage share. The nations may agree to pay percentages of the combined costs. If these add up to 100 percent, and are fixed, then the two nations would separately and collectively seek the minimum combined cost ("optimum") rate of TFP.

In Figure 3, the dashed line illustrates the second principle, sharing abatement costs. The distance above it is the share of combined cost borne by A, the distance below it by B. A minimises its costs by moving to the right, B by moving to the left, somewhere between zero and q_A^0. The dotted line divides combined costs in half, as suggested by the fourth principle. Both would have an incentive to move to q_A^0.

The "optimum" amounts of TFP referred to in the last two principles listed above are shown as q_A^0 in Figure 3. Muraro has shown that, on certain assumptions, this point will lie on the bargaining locus NQRP derived in Figure 2. Thus certain bargaining situations (all rights to one nation) and certain cost sharing principles (fixed percentages of combined costs) will lead to the same TFP, which may well be "optimum".

The pursuit of combined goals

Instead of bargaining or cost sharing, it is possible to regard the two countries as combining to achieve common goals. Instead of the conflict presumed in the "bargaining" and "CSA" sections above, cooperation and harmony may prevail.

Although the aims of such partnerships may of course run counter to world welfare (as will be shown) it will be generally assumed in what follows that the combined goals of neighbouring countries will also be the goals of the world as a whole. As the economist puts it, it is generally assumed that economic and environmental partnership to deal with one river's TFP will not generate excessive external costs for the other regions not naturally included in a pollution agreement.

We may classify the possible joint aims of the two or more countries under three headings: environmental, allocational, and distributional.

i) Environmental Aims. These may surely, in the 1970's, be regarded as independent of the other aims. All countries seek a cleaner global environment, cleaner water, an ecological balance that is closer to that found in a state of nature, and an ecology that is resilient and stable. Such aims are set forth in the declarations of OECD and the Stockholm Conference, and are not to be regarded as a mere re-statement of an economic demand for water quality acceptable for agricultural, industrial and touristic use. In economic jargon, the utility function for environmental quality may be taken as separable from the more direct enjoyment of cleaner water.

To say this is to give proper recognition to the world-wide environmental movement, and to the motives that have led to the OECD's concern about ecology and TFP. Thus countries may be willing to sacrifice both national and common benefits from better allocation of resources and distribution of wealth in order to achieve environmental improvements valued for their own sakes.

ii) Allocational Aims. It is not difficult to look at economic efficiency from the viewpoint of each country. But most economists will also wish to examine the "global" or "cosmopolitan" efficiency of the allocational arrangements. Furthermore, it is necessary to recall that each type of institution carries with it costs of transactions, information, agreement, administration, and so forth.

iii) Distributional Aims. Of course, as with most of international relations, TFP negotiations are dominated by distributional questions: who is suffering, who will benefit, who must pay? In the present context, we must keep in mind both questions of international abstract justice toward weak or developing nations, and practical questions of acceptable sharing among neighbours of costly projects. These aims however are not surveyed at this stage, but emerge in later discussions of rights, duties, and principles.

The efficiency aims of each country are easy to state. In dealing with TFP, each will wish to divert resources from their alternative domestic uses to the extent that the pay-off in protected environment, increased enjoyment, obeying international law, good relations with neighbours, or development of industry just equals, at the margin of transferring resources, their alternative domestic benefit. Given the rules of negotiation, this criterion will indicate how much they will be willing to spend on abatement and compensation. (Furthermore, if the rules are not already given, the criterion above indicates that national economic efficiency in the allocation of resources will justify allocating labour and capital to strategies that will in the long run secure the adoption of the most favourable rules. Thus national well-being may call for the obstruction or delay of agreement.) These aims are already familiar. Subject to qualifications about their lack of information about, or control over, their own subjects, it may be stated that the countries involved in the "bargaining" and "CSA" approaches reviewed earlier were seeking that allocation of resources (to abatement and to mitigating damages) that would minimise their total pollution costs. In the partial-equilibrium context of this discussion, to minimise these costs is the same as to maximise net benefits. Thus the "bargaining" and "CSA" approaches already contemplate allocative efficiency in the independent actions and responses of each nation.

However, these sets of actions may not be efficient from a combined, or global point of view. The national allocations of resources leading to, or arising from, TFP, may lead to a situation which is less than ideal from a combined, "global" or cosmopolitan point of view.

In what senses may, say, a stable "bargaining" procedure be effective?

First, the frontier. It may be that in arriving at a "bargaining" solution certain locations of industries and residences have been taken as given, while a "cosmopolitan" solution would regard them as movable. Thus, to take an often-cited illustration, it may be that the combined costs (as measured by the vertical distance between TC_A and GG'_B) could be reduced even below those at q_A^0 by shifting some of the polluters downstream from A and some of the victims upstream from B. Or, to take another illustration, it may be that the least cost TFP abatement technique would involve more dilution in B and less processing in A. (1)

---

1) In Figures 2 and 3, the curve GG'_B does actually contemplate abatement in the stream in B. See the text above in the neighbourhood of footnote 1, dealing with "mass treatment" downstream.

Hence, if every person and government acted as though the frontier did not exist, the total cost of TFP, to everyone, would be less. This would be achieved by a shifting inwards of the TC_A and GG_B curves, and is of the category of improvement often called "X-efficiency".

A second difference between the sum of national costs and the combined costs is in the perception of cost and damage. If each person upstream acted as though the people downstream were members of his own society, his group's TC_A curve would be higher for each level of TFP. Similarly, if each person downstream felt himself to suffer the upstream dislocations required to abate TFP, his GG_B curve would be lower. Thus ideal allocation of resources from a "cosmopolitan" point of view ("if the frontier did not exist") may call for a different standard of TFP than if the two peoples were hostile or indifferent to one another.

But whether or not people can disregard the frontier is a question of fact, not of economic criteria. If they can, if they wish to break out of their nationalistic bounds and include the whole river valley, or even the totality of the two nations, within their welfare-maximising horizon, then the "cosmopolitan" point of view is the appropriate one. Otherwise, the simple sum of national welfares is the appropriate maximand for joint policy.

In the third place, the tactics of delay and obstruction, quite consistent with an ideal allocation of resources within each nation, are surely wasteful from their combined point of view. Indeed, the separate countries may deem it worth-while to devote resources to misinformation, making sure that eventual cooperation is based upon faulty data. From a "cosmopolitan" point of view, however, real costs incurred with the aim of loading more of the costs onto the other party (even though well-spent from a national point of view) should be regarded in relation to the minimum amount needed to obtain agreement, having regard not only to the eventual levels of abatement and damage costs but also the costs of information, monitoring, agreeing and private transactions.

To summarize, the "cosmopolitan" point of view does not call for the complete abandonment of outlays that have a nationalistic intent. It would be a mistake to disregard individual and collective tastes in each nation concerning such matters as the location of sewage and abatement works, the location of consumers and industries, and the extent of redistribution implicit in alternative bargains. Just as national economic efficiency (Pareto optimality) can be defined only in terms of a given distribution of income, so international efficiency must also be defined in acceptance of given constraints on the possibilities of re-allocating, or relocating, resources between nations.

Hence, the definition of international or cosmopolitan efficiency in the design of TFP institutions cannot be insensitive to the norms and aspirations of each party. For experts to design an agreement that will work out to make the best use of resources "as though no border existed" is to achieve something that neither party seeks. The border is there, and may symbolize a discontinuity, real or desired, in tastes, incomes, relative prices, land use, role of government, language, infrastructure, and so forth. (1)

As already suggested, in our discussion of CSA principles, the "optimum" "cosmopolitan" TFP may be visualised on a diagram such as Figure 3. Here, in Figure 4, we refer to welfare contours. Instead of regarding TC_A as a cost curve, each point on it measures, vertically, a payment to A at the level of TFP measured horizontally. On TC_A, of course, these payments will just cover A's costs. Hence A is indifferent between all points on TC_A; its costs are just covered; it is everywhere just as well off as at q_A. This curve may therefore be seen as the highest curve of a set of "indifference curves" for A. The curves drawn below TC_A represent, successively, smaller payments to, or lower welfare, for A.

---

1) We must avoid Harry Johnson's (Senate of Canada, 1971) disparagement of "economists, whose job is to study countries as cases, without any real demonstration that a country is an economic unit that is worth talking about." Even if each country cannot escape from international prices and interest rates, it may wish to, and be able to achieve a different allocation among private, public and environmental goods, and even a different redistribution of income. To repeat, these are matters of fact, not doctrine.

Similarly, we may regard $GG_B^!$ as a series of points on which payments are received that just cover B's costs from that amount of TFP. The other curves above GG_B represent smaller payments to B (or larger payments from B) all payments are measured vertically upward or downward from the horizontal axis q_A.

If the curved lines for the four typical CSA principles were not drawn in (as in Figure 3; not shown in Figure 4), it would be seen that the PPP principle (any point on the q_A axis) would cut A' highest indifference curve at q_A^1, and that B's highest accessible welfare curve would be at zero. Similarly, the fixed-sharing-of-combined-cost principle would be tangent to both sets of indifference curves close to a lower TFP, which we have previously labelled the "optimum".

For the present discussion, we may eschew these nationalistic CSA principles and instead note that, from a "cosmopolitan" point of view (see Figure 4), each country's iso-revenue curves will be tangent to those of the other along the locus which we previously named NQRP. This locus is analogous to a contract line, along which allocation of resources to TFP is "Pareto-optimal", given the production functions, factor endowments, tastes and pollution rights of the two parties. (1) Any position off this line is inferior, for both countries, to some position on the curve.

Understanding this diagram makes it possible to emphasize three points that are not so obvious without the use of geometry.

First, achieving the optimum degree of pollution almost certainly involves a payment or transfer. This conclusion has emerged gradually, and has not been stressed. But, unless by accident the parties arrive on a bargain on the q_A axis, some payment must be made.

Second, unless the curves are parallel, there is no unique optimum degree of pollution. The optimum depends on the initial distribution of welfare and payments. For example, if the initial situation is at q_A^1, the "social optimum" for the two countries combined must be between N and Q (Figure 4). Any lower point reduces A's welfare. A similar example: if the initial position finds B at zero TFP, then the social optimum is at the bottom end of the contract line. A third example: if it is agreed that neither party is to make a cash payment to the other, (i.e. the final point must be on the axis), the optimum is at q_A^O, where the contract line cuts the axis. These three initial distributions may yield three quite different "optimum" levels of TFP.

Third, the "optimum" level of TFP is unlikely to be near the extremes (zero or full) that would be chosen by either country. As has been suggested by Dolbear for externalities in general, the optimum amount of TFP, assuming its production is of benefit to one person and harms the other, lies between the extremes. (2) Thus any compromise point between complete pollution and complete purity is likely to be closer to the "global optimum" than the extreme that either would choose.

This is almost all that can be said about the "ideal" or "optimum" pollution from a global point of view. The indifference curves can be varied somewhat from their position when purely nationalistic attitudes were represented, but it is unlikely that their shape will change drastically. The resulting "optimum" is only a modest step towards the best of all worlds, environmentally speaking: it is

---

1) If the curves are interpreted like iso-revenue curves, they can be drawn parallel to TC_A and GG_B and so will yield a vertical contract line. They are simply combinations of TFP in which social benefit or cost are one dollar higher then on the base curves. Alternatively, they can be treated as derived indifference curves. If TFP is unimportant in relation to each country's GNP, then the underlying indifference curves are parallel. Hence, if there are constant returns in water-using industries, etc., the curves here will be parallel, again producing a vertical contract line. However, if a country is poor, or if TFP looms large in relation to GNP, then its curves will not be parallel, the contract curve will wiggle, and there is no unique optimum TFP.

2) H. Shibata, "A Bargaining Model of the Pure Theory of Public Expenditure", Journal of Political Economy, Vol. 79 (1971), pages 1-29; and F.T. Dolbear Jr., "On the Theory of Optimum Externality", AER, Vol. 57 (1967), pages 90-103.

simply a particular combination of payments and TFP that makes both nations "better off" than initially. But who is to say that in a world of extreme pollution and variability of wealth, the <u>initial</u> situation is in any way a sensible point for comparison? Certainly not the economist.

Reciprocal Pollution: Boundary Waters

The discussion so far has centred on upstream-downstream pollution. In this section the model is brought to bear on reciprocal pollution, such as might exist across a boundary lake, river or sea. Both predictive and joint-action analyses are used. The OECD literature rightly points out that, compared to one-way pollution, a wider variety of solutions is available for TFP of boundary waters, because the parties have both similar rights and similar damages.

We begin with the questions raised in page 181 above for one-way pollution. The first of these was the prediction of the independent actions of the two countries. This question may be regarded as an extension of our question about the behaviour of a single downstream country to deal with two countries, each confronted with the discharges of the other into a common lake.

The second question was the nature of the joint action, and its optimalty.

Independent action and reaction

The assumptions for these questions are as follows:

i) Public good assumption: any given amount of pollution (or abatement) is experienced (or enjoyed) equally by both countries.

ii) The amounts of abatement produced by the two countries are additive.

iii) The countries are initially assumed to have the same abatement cost functions. This is a very strong assumption; when one considers the various ways in which pollution may be reduced, the assumption is tantamount to assuming identical economies.

iv) They are also assumed to have the same damage functions.

v) Marginal costs. The results to be demonstrated depend on the assumptions about <u>marginal</u> costs. Unlike the discussion thus far, both marginal costs are assumed to be rising at an increasing rate.

To commence our examination of independent reaction to reciprocal pollution, we picture a lake on which only one of the two riparian countries has settled, A for example.

In these circumstances, A will minimise its own costs (AC + DC), as shown in the first panel of Figure 1. It need not consider the other country. Marginal abatement costs will equal marginal damage cost (mac = mdc) at q_A^i as shown in Figure 8. The identical cost assumptions mean that q_A^i must equal q_B^i, if B were the only developed economy.

When country B arrives, and proceeds to develop to the same level as A, a new situation emerges. Various sequences are possible. For convenience, assume that B initially pays no attention to pollution, and copies A's setup in every respect. Then the discharge into the lake will be equal, and will mix, so that following assumptions (i) and (ii), both experience the same amount of pollution. Thus until new decisions are made, each of the identical countries now experiences twice as much pollution as A did before the coming of B.

Then the behaviour of each becomes that which was described by B's reaction curve in panel 2 of Figure 2; B takes A's TFP as given and adjusts its own emission of pollution so as to minimise its total costs as a function of q_B plus q_A. Two of these curves are shown as Figure 5. They express the assumption that each takes the other's pollution as given.

If either country had the lake to itself, its pollution discharge would be q^i. But when both suffer the other's discharge, they converge on the smaller amount q^r, the coordinate of the stable equilibrium point E. The total amount of pollution shown by the dotted contour lines, is larger than when only one country made a net discharge into the given volume of water, but less than twice that amount.

191

That their minimisation of total pollution costs causes such a reduction in country B's discharges was shown in Panel 2 of Figure 1. Every increase in q_A^1 causes a fall in q_B^1. Since this reaction takes place in both countries, the final amount of pollution is less than double the original.

This can be shown more precisely with the marginal costs of Figure 8. Recalling that the two countries' mac and mdc curves are identical, it follows that in the final equilibrium for which we are looking each will have made the same adjustment. By assumption (i) each suffers from its own and the other's pollution. Therefore, in equilibrium its marginal damages for each amount of its own pollution discharge will be equal to the marginal damage for twice that amount of its own discharge in the absence of foreign pollution. This is shown in Figure 8's mdc_{A+B}^* . It is constructed to bisect the horizontal distance between mdc_A and the vertical axis; thus it shows the mdc experienced by A when both A and B produce the amount of pollution measured on the abcissa.

Thus under assumptions of independent behaviour, reciprocal pollution causes the discharges of A to decline from q_A^1 to q_r (i.e., from E to G).

The greater the rate at which the slope of mdc increases, the larger will be the vertical distance between the two curves and, therefore, the larger will be the reduction in A's discharges. If, however, as may happen, marginal damage costs are fixed (horizontal) in the neighborhood of E, G will coincide with E. That is, as common sense would have foretold, if people in A do not mind a lot more pollution, they will make no extra effort to abate their own discharges. (Other possible slopes for both mdc and mac are considered below).

Joint action

We now compare this outcome with the pollution that would be produced under joint action by the two countries. Joint action would emerge if the two economies and nations merged completely, and acted thenceforth as one state to maximise national welfare. But such drastic assumptions are not needed. We need only imagine that it is decided, in the case of this particular river, to ignore the boundary and to abate pollution to the extent that would be economically efficient. Thus A and B would jointly minimise their total costs of pollution, combining the same curves that have so far motivated their independent national actions. One test of success if this endeavour would be the extent to which they had succeeded in equating the marginal costs of damage and abatement everywhere, even if the curves were not identical.

The relevance of this to our essay is obvious: joint action requires institutions, while independent action does not. To determine the need for joint action, three questions must be answered: would the solution be different from that already outlined for independent action; would it be more efficient; and would it be worth its transactions and bargaining costs? Herein we assume that it would be more efficient; we will not consider those types of joint action in which new special-purpose institutions might autonomously decide to act less efficiently than its two parents wish. We leave to the second half of the essay a discussion of some of the contracting and bargaining costs that might arise. In what follows therefore we deal only with the contention that the pollution that would be produced under the "independent" conditions sketched above is more than would be produced by joint action. To predict that decision, using the same marginal cost functions, we need only double each cost curve. In this way the diagram that has so far described the behaviour of half the pair of identical countries may now be used to compare the total behaviour of the two parts under independent and joint actions.

Consider mac_{A+B}. It represents the costs of abating in the two independent countries, each trying to minimise its total abatement costs. It is constructed so as to double the horizontal distance of mac_A from the vertical axis. Since mdc_A also doubles the distance of mdc_{A+B}^* from that axis, it follows that intersection F has the same cost as intersection G and represents twice the amount of pollution - i.e. the sum of the amounts that the two countries produced independently. $(q_{AB}^{2r} = 2q_A^r)$.

Now consider joint action. The combined abatement cost is still represented by mac_{A+B}. But there is a new marginal damage cost function. Whereas mdc_{A+B} represented the marginal damage suffered by one country on the assumption that both countries produced the same amounts of pollution, mdc_{A+B} represents a vertical doubling of mdc_A. That is, it shows the total marginal damage when the pollution is treated as a joint public bad, and the abatement as a joint public good. The total cost minimisation decision of the joint jurisdiction would equate mdc_{A+B} with mac_{A+B}, at H.

Thus to compare independent and joint action, we need only compare H and F. Because both lie on the same combined marginal abatement cost curve, one must definitely lie to the left of the other. And since by construction mdc_{A+B} lies above (to the left) of mdc_A, H must lie to the left of F. That is, joint action will lead to less lake pollution than the sum of independent actions.

This conclusion may now be exposed to the relaxation of the assumptions on which it is based. First, consider the shape of the marginal cost curves. If both are constant (horizontal), there will be no unique equilibrium. If abatement costs only are constant H will lie on a line connecting G and F: the conclusion is unchanged. However, if damage costs only are constant (over the relevant range), H and F coincide: this is a limiting case. We shall not consider cases where the marginal cost curves have the opposite slopes from those assumed. If it is assumed that the cost curves slope upward at a constant rate (instead of rising at an increasing rate as illustrated), the two mdc curves will coincide but H will still lie to the left of F. Thus the main conclusion seems fairly insensitive to the various second-order conditions that might be encountered. Next, consider assumptions (iii) and (iv), that the two countries' abatement and damage cost curves are the same. They invite disbelief. Abatement costs are thoroughly embedded in the productive processes of the economy thus identical structures of industry are necessary to obtain an identity of abatement costs. It is just as hard to believe that damage cost curves, depending on wealth and population as well as on productive processes, could be the same. Can we generalise about the consequences of a relaxation of these assumptions for the joint-action conclusion?

If the damage-cost curves are different, the only generalization that seems to emerge concerns cost differences. For example, abatement cost functions might be the same but A would suffer greater marginal damages at each level of q. Then because independently A would abate more, it would run into higher marginal abatement costs than would B. Under joint action, some of A's desired abatement could be undertaken using some of B's unused abatement capacity, at lower marginal costs. The equalization of marginal abatement costs would permit more abatement than under independent action. (This conclusion of course depends upon upward-sloping marginal-cost curves). Or, take the opposite example: both countries have the same damage functions but B now has higher marginal abatement costs at all levels. Now B would be expected, independently, to abate less than A. But under joint action, some of B's desired abatement could be undertaken in A, until marginal costs were equal.

Thus, both examples suggest that joint action, providing an opportunity for one country to "export" abatement services to the other, will lead to more abatement taking place when marginal costs are equalised than when the two countries act independently. This conclusion tends to act in the same direction as that already reached. Whether or not there are identical cost functions, joint action will lead to less pollution and more abatement than independent action.

Notice, however, that "exports" require international repayment. Unless the two countries have merged completely, one will abate for the second only if it is paid to do so (in cash or in kind). Thus, as with one-way pollution, the optimal adjustment to reciprocal pollution will involve payments.

Strategic models

Of course, a number of other models of two-nation action and reaction are possibly applicable. One way of arriving at the "independent" results above is to assume that a process similar to that in the Cournot duopoly model is followed. Each nation would act as though its neighbour's pollution were fixed; as though it were part of the "background" quality of the water. But continued action and reaction should soon falsify such beliefs, and lead to the joint action also examined above.

Instead, we could have recourse to models involving game theory and strategic behaviour. Each country realizes that the two countries are interdependent, and that both of them know it. How will they decide to play it? (1) Another set of models would involve the concept of "leader" - a nation A that dominates the use of the common lake, expects the other nation, B, to act passively, and so sets a policy that attempts to reduce both B's costs and A's costs below their levels if they did not play their leader-and-follower roles. Of course, from such role-playing it is only one step further to collusion, under which it is likely that the leader-and-follower roles will be resumed within the privacy of a common council or agency. Ordinary collusion, as in price-fixing, is unpopular because it leads to inefficiency and loss of consumer's surplus. But international collusion may be more meritorious, leading to lower abatement costs, less pollution, and more goods.

There is also room for speculation, here, about whether an increase in the number of separate decision maker (countries) will bring the Cournot equilibrium closer to, or farther from, the optimum for them collectively. Oligopoly theory would suggest that as numbers increase, the optimum is approached. As the number of neighbours increases, the changed action of any one has a steadily smaller effect on the situation felt by the others, so that each is increasingly justified in taking some independent action.

On the other hand, the lack of information, and uncertainty, may increase. Each country may feel that independent action is stupid, with so many, unpredictable, neighbours. Then an n-country situation may, in the absence of joint action, result in the sort of chaos that would be experienced on roads and bridges if there were no agreed traffic authority.

There is, at this stage, need for far greater study of the n-country type of model. It obviously applies to some lakes and seas, and to the oceans as a whole. But the subject has not yet been tackled, as far as we know; ocean pollution from rivers and estuaries, for example, is rarely discussed by economists. In this survey, therefore, we confine ourselves to a two-country model.

Is joint agreement necessary?

The outcomes of the two-country model so far, are rather startling in their application to present joint-TFP situations. They tell us (1) that countries will abate separately even if they do not cooperate, (2) that if the countries are identical, they still have need for joint action, (3) that if they do act jointly and are unlike, they must make transfers between them in order to reach a combined optimum (in other words, one country must abate more than another).

None of these conclusions seems to accord with our observation of international events and institutions. Why is this? The answers to these questions tell us something about the model's deficiencies, and even more about what we should look for in the real world.

First, we do not hear much of independent reaction in the absence of combined agreement. There are a number of possible, sufficient, explanations for this.

   i) Strategic behaviour. Neither country wishes to abate pollution independently for fear of prejudicing the concessions to be demanded from the other country at a later bargaining session.

  ii) Abatement is taking place. Neither country presses the matter, but each is already independently restricting the dumping of wastes on its shores.

 iii) Incomplete mixing. Explanation number 2 may be even better when we recognize that certain kinds of pollution do not mix completely across large lakes. Then local restrictions on pollution are certainly already in existence, but are not recognized as being also, in part, caused by TFP from elsewhere on the lake. (Examples: the Baltic, Montreal and the Province of Quebec, the upper Great Lakes). Any of these explanations is sufficient to explain the lack of discussion of independent, Cournot-type reaction. In the present context the first, strategic, explanation is particularly crucial. While the preliminary debate on TFP responsibility is proceeding, countries that might otherwise independently help to clean up

---

1) Cf. the discussions by Muraro, and Smets and especially R.C. D'Arge in Problems in Transfrontier Pollution, OECD, 1974.

lakes and seas, find it prudent not to do so for fear of suffering worse bargains later.

Second, we hear little enthusiasm for joint action. Fear of loss of sovereignty at the hands of nationals of the other country must be the chief reason. Joint action indeed may bring low abatement costs by requiring industrial relocation, and this may be dreaded rather than sought.

Third, we do not observe transfers or payments between unlike countries sharing a common lake. Yet both profess themselves made better off by some treaty or common program. Is this possible? Are today's agreements actually benefitting both parties? If not, why do they agree?

As we have seen, two unlike countries would both benefit if each independently abated pollution more than if they did not suffer from TFP from the other. We have suggested above that they will probably be found to be already doing so, prior to any agreement. Of course, if they do sign an agreement, they will continue to abate, as before the signing. However, the two countries would be more ready to act independently if they could be assured that the other would not wish to change its economic activities, or pollution, in the future. Thus they will carry out what should have been "independent" reactions more readily if they have a treaty to assure them that the other country will not later increase its pollution, or will abate it. That is, a treaty can confirm the expectations that are assumed in the Cournot model, and make them permanent.

Another reason for agreement without payment has to do with domestic politics. It may be administratively difficult for the larger country to proceed to benefit itself by abating pollution independently. Householders, internationally competitive firms and taxpayers at large will complain that their opposite numbers in the other lake-side country are not being asked to make the same "sacrifices". Thus the government of the larger country may enter into an agreement, costly to itself, simply to assure visible uniform behaviour. Similarly, the smaller country cannot persuade its most-affected citizens to conform to an advantageous independent abatement policy when the larger country across the frontier, apparently continues its habitual dumping of pollution.

Finally, of course payment (transfers) may take place. What may appear simply as agreed uniform (equal-percentage) reduction of pollution may not in fact require equally costly abatement. For example, if A's activities give it more opportunities to abate a certain pollutant at lower cost than B, A may be actually paying B in kind for the latter's action (that is, A specializes in abating the pollutant in which it has the low cost comparative advantage in return for B's specialization in abating another). Furthermore, payment may be taking place in kind in other goods. Who knows, for example, how many ways there are for Canada to reward the United States for apparently similar policies, when the main burden falls on the United States. Why does the United States agree? Of course, up to a certain amount of abatement, it experiences a net benefit, in many parts of the lakes. But, even more important, it has scores of ways in which it depends on Canada: water supply, water levels, navigation, highways, railroads, bridges, not to mention trade, migration and defence. Final agreements are made between the two chiefs of state, perhaps, because only they can perceive what the quid pro quo may be, for actions by one that increase the efficiency of both. (1)

Thus it can be seen that observation of the real world does not offer refutations that are dangerous to our model. It is likely however that further study of such models will cast even more surprising light on the actions of states around a lake. It does appear that efficiency may well be lost unless the countries specialize in reducing those pollutants for which their costs of abatement particularly suit them. Such "specialization" may, if there is more than one pollutant, be reciprocal. If it is not reciprocal, and if payments cannot be made, it is not clear why any agreement on TFP is contemplated. Independent reaction may well produce nearly as much abatement as combined action without payment.

---

1) This explanation may also apply to the Swiss-German-Austrian treaty for the protection of Lake Constance, of 1960. In any case, it is interesting that this treaty, and others like it, often call for the countries simply to pursue their already-existing domestic policies effectively. That is, by the treaty, each is simply being given the reassurances that would justify independent action.

Conclusions on model

Nothing can come out of the model that was not put into it. In this case, the results stem mostly from inserting into the models some conventional assumptions about increasing marginal costs and decreasing marginal benefits of pollution abatement (i. e. increasing marginal damage of pollution), and the slightly less orthodox assumption that each country can be regarded as "passively" minimising its total costs (maximizing economic welfare) with respect to AC and DC in response to each TFP decision or transfer made by the second country. There are other assumptions - about strategic and diplomatic behaviour - that could have been made, but they usually apply to short run advantage rather than permanent settlement. But it should be noted that our assumption, which implies that senior and junior governments make endogenous, optimal, market-like, decisions, plainly requires more than the usual individualistic underpinnings of most international trade or environmental applied theory.

It has also been assumed, implicitly, that all damages and abatement costs can be converted into a common numeraire. The countries' valuations of this numeraire (their "marginal utility of money") may differ, but the numeraire is a perfectly standard alternative to TFP.

To what conclusion do these assumptions bring us?

First, that neither party is likely to seek to reach a middle, or an optimum, level of TFP. The upstream party benefits most by a high level, the downstream party by zero, TFP.

Second, if either party has "rights" over the stream, the optimum or middle level will be agreed upon only if that party is recompensed by payments or transfers.

Third, to be required to accept a middle, or an optimal, level of TFP without compensation involves a considerable transfer of wealth between the two countries.

Fourth, TFP on a shared lake will be abated to a certain extent even if there were no treaty or agreement. The "optimum" abatement, however, may call for payments or transfers, just as with upstream-downstream pollution.

INSTITUTIONS

Introduction

In this second part of the paper, I turn to the choice of institutions. By "institution" is meant both the sequence of procedures to be used by politicians, administrative officers and negotiators, and the organizations within which data is received and processed, coordination takes place, decisions are made, appeals are heard, and long run and day-to-day "management" of TFP services are carried out. What does the "management" of TFP services include? Among other things, the administration of central sewage and abatement works; stream and pollution measurement and monitoring; prescription and enforcement of private abatement activities; handling inter-jurisdictional financial payments; and attempting to reduce the vulnerability (by relocation, etc.) of persons and firms to damages.

Three points should be made. First, it is not usually the task of some international agency to undertake all these roles. It is to be expected that most of them will continue to be the responsibility of present national, provincial or local organisations. One of the tasks for the economic study of international institutions is to reduce to the optimum the cost of over-lapping or redundant agencies and branches of government. Consequently, the OECD should hesitate before recommending the superimposition of new institutional layers, especially layers of operating agencies.

Second, it is unlikely that the style or system of bodies on opposite sides of the frontier will fit well together. (Engineering or public health agencies may not have difficulty in cooperating or even merging; but financial, social and political bodies may find joint action very unfamiliar.) Consequently there will always be a need for some special institutions simply to transmit and translate data and intentions, or to prevent friction developing when the abatement mechanisms of the nations are meshed together.

Third, there is no need for TFP institutions to be monolithic. Just as river administration within each country is very often efficiently shared among national, provincial and local bodies, so there may well develop a mosaic of international institutions. The nature of these "institutions" may range from joint managing agencies through informal coordinating committees of local administrators to infrequent high-level meetings of national leaders or experts. Such international institutions, because they are all concerned with aspects of the same question, must at least be aware of each other. But it is not obvious that they must all be part of a single organisation, or under unified, continuous, control.

Agenda: Typical sequence of international action of TFP

In the following pages we will survey possible arrangements for dealing with TFP as though there were no previous contact or liaison between the countries involved. This assumption is only for expository convenience. In reality, many problems of river navigation, public roads and bridges, customs and passport administration, irrigation and water supply, hydro-electric generation, flood notification and prevention, fisheries and wildlife protection will already have accustomed various national agencies and officials to joint action and consultation within the river basin. Certain procedures will already be institutionalized. In addition there will also be a "trans-national" structure of non-governmental links and transactions which should be recognised and preserved wherever useful for TFP cooperation. These existing institutions should not be lightly abandoned; they probably form a firm basis on which TFP institutions and procedures can be built.

Allowing for the fact, therefore, that many institutions will already be in use, we may visualize the sequence of joint action on TFP as following on worldwide discussion of TFP problems (oceanic oil spills; fall-out and atmospheric pollution; legal discussion of rights to international river levels and flows) and local complaints about increasingly serious TFP or about particular spills and incidents. It is assumed below that this discussion, reflected in the OECD and at the IMCO and Stockholm Conferences, has already taken place. That is, there is no need to design institutions to stimulate discussion of and demand for TFP abatement. Thereafter, the most important areas in which institutions will be engaged are the preliminary studies and the bargaining alternatives.

Preliminary Studies

Before negotiations can seriously commence, both parties must assemble, separately and together, the raw material which their later sessions will be able to fashion into a treaty or agreement.

Technical Information

It may seem obvious that information and data are necessary before an agreement can be reached. What is not realized is that this is a very slow and demanding process.

In the first place, merely detecting some pollutants, let alone measuring their flow and density, calls for decisions about the timing and placing of sampling stations. These decisions must take into account the need for data which both countries can accept which is not unduly influenced by unusual events such as floods, high water, drought, spills, landslides, road constructions and fill, land-clearing, failure of new sewage or abatement equipment, or change in river channels. Thus several years' records, at least, will be required.

In the second place, in addition to the seasonal levels and flows data mentioned above, it is necessary to have an understanding of the hydrology of water bodies near the frontier, especially in lakes. Otherwise later discussions will be threatened with disputes about the points of discharge of certain pollutants.

Third, it is necessary to compile records or estimates of emissions at each point on the river or lake, to compare with measurements of pollution densities and hydrology.

Fourth, in addition to these types of flow information, it is necessary to estimate stocks: of chemicals and debris and waste build-ups in the water or in beds of streams and lakes, not to mention stocks of fish, shore-birds and animals.

Fifth, it is also necessary to establish the correspondence of the national systems so that each nation will be able to understand what it is negotiating, how the TFP situation compares with its domestic pollution problems, and how it must continue to collect data that is informative to its partners.

Economic and social information

Economists will jump to the conclusion that what is required is knowledge of the damage cost and abatement cost functions mentioned earlier. Such estimates will indeed be useful. But even more important in the long run will be the collection of raw social and economic data which can be used not only for such estimates but for gaining more direct and elementary understanding of the shared river or lake as (i) a waste-disposal facility, (ii) a source of water supply, (iii) a fishing resource, (iv) a means of navigation, (v) a means of drainage, (vi) a recreational and landscape amenity, for farms, towns, industries, ships and resorts.

To gain this information may require the countries to redistrict their censuses so as to exclude population, occupational, industrial and trade data on locations which are near to, but not within, the basin in question.

At the same time, it will be desirable to learn the extent to which other parts of each economy are dependent on the river or lake for the five functions listed just above. (This is easy to say, but "linkage" or "dependency" information is notoriously difficult to obtain and interpret).

Damage Costs

The model developed in Part I of this paper has shown that in an upstream-downstream TFP situation, efficiency and equity both require a common understanding of the damage-cost function. That is, agreement will be promoted by accurate, accepted measurement of the amount of damage caused by each level of TFP. This measurement will be used, by both parties, as raw material to determine to what extent pollution should be abated, and how the total costs of the final situation should be shared.

I am very pessimistic that much progress can be made in measuring the damage to the required extent. While measurement of the present damage (that is, before TFP is abated) will be of some help in bringing about agreement about the need for some action, this measurement is a very poor indicator of the damage at lower levels of TFP.

The easiest damage information to obtain is derived directly from the extra costs incurred by industrial and municipal water users (compared to similar users elsewhere) in dealing with pollutants in their supplies. Estimates might also be made, fairly accurately, of the gain to fishermen and farmers.

But when pollution causes damage to the enjoyment of non-marketed services (tourism, values of view to residences and offices, bathing, sport fishing and hunting, boating), the damage caused must be measured extremely indirectly, by methods pioneered by Clawson, Knetsch and others.

Even worse, many of the benefits of pollution abatement are experienced as public goods. Only Peter Bohm has any confidence that it is possible to measure, by questionnaire methods, such values.

Finally, it must be stressed that whatever methods are used to obtain the above assessments, they will almost certainly apply only in the short run. That is, they will not indicate the diminished damage cost after enough time has elapsed for there to be a complete cost minimising adjustment to a new TFP level. If, for example, TFP were to be reduced, the "adjustments" in a downstream country would, in the long run, include the modification of products and services using water as an input; change in the scale of filtration equipment; moving of recreation sites, resorts, restaurants, etc., as well as water-using industries and agriculture; new requirements for downstream abatement; and new expenses caused by changes in taste resulting from the new opportunities to experience improved environmental quality. These changes in total damage cost level for a given TFP will emerge only in the long run. Although the effects may offset one another, examination of them does suggest that long run damages will be lower, perhaps considerably lower, than whatever level appears in the short run. (1)

To summarize this brief discussion of estimates of damage cost, it may be predicted that countries will be forced to get along with more primitive indications of the benefits of TFP abatement. While the costs of a few types of damage can be measured, most cannot. The most insurmountable difficulties occur in measuring DC connected with the production or enjoyment of non-marketed goods and services, and (even more worrisome) in the production or enjoyment of public goods. Furthermore, any acceptable estimates are likely to exaggerate somewhat the long run damages, because of the long period required for the victim economy to adjust itself efficiently to a changed level of TFP.

Nevertheless, bargaining will be almost impossible unless the country that is suspected of causing TFP damage can be assured that harm is being done, or that costs to mitigate the harm are being incurred. Thus it follows that (however unsophisticated and refutable the indicators of downstream damage) the estimates should be prepared by experts drawn from or representing both countries so that subsequent discussions benefit from scientific, sociological and economic criticism rather than suffer from inflamed nationalist partisanship.

Abatement Costs

The earlier model will also lead to a demand for a measure of abatement costs. Here we are on much more solid ground. Given time and tolerance, specialists can estimate the cost involved in abating each kind of pollution to a given extent.

To recapitulate an earlier definition, these include the alternative costs of re-cycling wastes; of disposing of wastes in another mode; of reducing waste-emitting production processes; and of changing location. To a great extent, these costs, especially in the short run, will consist of the sum of the expenses of pollution-abating equipment and structures and the loss of income from processes curtailed. In the long run, these initial outlays may be mitigated by turning to substitute production methods, raw materials and final products, and by changing location of production.

This is not the place to discuss cost estimation: the literature (comprehending writings in sanitary engineering and public health) is very broad. Two observations are however essential.

First, as with damage costs, abatement costs may be experienced elsewhere than in the region where the TFP abatement takes place. They may take the form of increased land, water or congestion costs elsewhere. Furthermore, they may take the form of changes in general economic development for the whole economy. Indeed, much of the discussion about the problems of international environmental protection stem from the likelihood that abatement costs will either syphon off investment in less developed countries into abatement outlays, or deter investment altogether.

This is not academic speculation. In many low-income countries the most attractive industrial or urban sites are on a river or lake upstream of other nations. In such societies abatement costs are seen as additional difficulties in the way of industrial development that would, if undeterred, have linkages throughout the entire economy. Thus abatement costs are not perceived simply as static opportunity costs of a slightly different allocation of resources, but the threatened sacrifice of forgone or postponed development of an entire society.

In slightly more advanced economies, abatement costs may be revealed less as forgone development than as sources of increased costs of production that will

---

1) This statement requires two amplifications. The magnification of short run over long run damage costs also depends on whether the change is toward more, or less, TFP. However, the long run level will always be below the short run. Note that, in the long run, some of the damage costs of a changed TFP level may not be experienced within the river basin, but as changes in congestion, land values, or amenity in adjoining regions.

imperil a nation's capacity to sell already established staples. In a nutshell, a nation depending on the export of goods that create downstream pollution in the course of their production, can abate this pollution only by a deterioration in its overall terms of trade; that is, in its standard of living or rate of growth. (1)

Therefore, to approximate abatement costs of TFP in terms of money expenses or even static opportunity costs would be wrong for those countries where actually abating discharges would have a more profound effect on the whole economy. What is required for subsequent bargaining is knowledge of these effects.

This brings us to the second observation. Just as with damage costs, it is essential that (however unsophisticated and refutable the indicators of abatement costs for each level of TFP) the estimates should be prepared by experts drawn from or representing both countries. This will enable subsequent discussion to deal constructively with the methods of calculation and their probable accuracy, rather than with partisan claims attacking or defending estimates made by one country only.

Optimizing combined costs

When the two countries are considered together, their combined costs, both of abatement and of damages, may be smaller than the sum of their separate costs. This is because the best way of dealing with ("adjusting to") TFP may be, in the long run, relocation of both sources and victims of TFP. For example, if the two countries contemplate agreeing not only on TFP but also on the migration of workers, trade, investment flows, or similar matters, the experts preparing information should investigate the possibilities of making improved international use of the river basin.

The assimilative capacity of the river, for example, is always best used if the distance between sources and victims is increased. This may best be achieved by moving certain industries far upstream. Just as plausible is the possibility that dependence on assimilative capacity may be reduced by reversing the above proposition, and placing pollution sources downstream of victims. This may best be achieved by moving certain industries far downstream, and the victims upstream. (2) In either case, the experts will have to report not only on the technical TFP-damage possibilities of such industrial migration, but on the social and economic consequences of location for the firms and people involved, and for their two nations. Sometimes the steps investigated may be minor: an abatement plant might be located across the frontier from the nation whose effluents are being treated; fishermen might be allowed to share the catch of the most pollution-free portion of a boundary lake or river regardless of the boundary; or picnickers from both nations might be welcomed to an expanded park and beach site on a specific stretch of the boundary-water shore. Sometimes the steps might be more adventurous, having major effects on population, GNP, tax revenue, and so forth. Unless the preparatory information of such possibilities is produced, it is unlikely to emerge in the less enterprising atmosphere of subsequent legal and diplomatic bargaining.

Once again it should be stressed that the reports on such integrated river-basin usage should be prepared for the subsequent meetings by joint groups of experts. Discussion of the possibilities should center on their desirability and on the distribution of their cost and benefit, not on their correctness or possible bias.

---

1) However, developing economies may well be equipped with more modern productive processes, and abatement techniques, than already developed economies. Hence abatement, measured in product terms, will be less costly, Cf. C. S. Russel and Hans H. Landsberg, "International Environmental Problems - A Taxonomy", Science, 172 (25 June 1971), pages 1307-1314.

2) In effect, the assimilative capacity of the river is being augmented by that of the next receptor downstream, probably either another country or the ocean. It should not of course be assumed that the two countries with which we are dealing are free to pass their combined pollution to this next region.

Hearings

At this early stage, information should be obtained from the people to be affected by the change in damage or cost. The normal way to do this is through the political channels of local, provincial and national governments, and these should be fully used. Debates, as early as possible, if based on full disclosure of what is being investigated can help not only to crystalize the issues but also to provide new ideas and information. It should always be remembered that the experts employed on the preliminary studies do not know everything, and cannot think of every possibility. Political discussion, properly encouraged, can expand knowledge.

The same is true of public hearings. In many countries today it is impossible to proceed with major changes in the social or natural environment without providing an opportunity for affected groups and individuals to offer their opinions; it may be assumed that this is bound to take place. Experts often regard such proceedings as at best providing a costly safety-valve through which pent-up feelings can be released without social or political damage. This is a limited view. Before hearings are staged, the public should be provided with literature making clear what is known and what is yet to be learned, before any political or international decisions or bargains can be initiated; then the public can be invited to offer relevant opinions on the facts (damage, abatement costs, opportunities for river use, preferences about institutions, and so forth). These proceedings can be time-consuming and may lead to public expectations about subsequent events that are unrealizable. But if carefully explained and carried out, they can, like political discussions, lead to an increase in knowledge both on the part of the public and the press, but, just as important, on the part of the experts responsible for the joint preliminary studies.

Disclosure

In the paragraphs above, the need for joint investigation of the effects of TFP and its abatement has been stressed. The reason given has been that if partisanship is squeezed out of the process of data collection, bargainers can concentrate on substantive (political) questions. (1)

The same argument can be put another way. If data is collected by national organizations, it will be under the jurisdiction, in part, of the very people whose duty it may be to represent their country in subsequent bargaining sessions. Because it will subsequently be their duty to win the best possible advantage for their country with respect to TFP and its removal, it will naturally occur to them that part of this struggle can be won by influencing the information that is disclosed to the other country. Certain estimates will not be made, or not presented, and certain other figures will be emphasized, or even exaggerated. Each country will know that the other is acting in this (perfectly natural) way, and each will go to the sessions prepared to attack the other's figures, rather than to work constructively. (2)

Two results can be predicted. Progress will be slow, because the data base cannot be agreed upon. And, perhaps more serious, the final agreement will select joint policies that either make use of that portion of the disclosed data that both can accept, or that require no data disclosure for their subsequent administration. Both results may be costly to both countries, compared to the characteristics of agreement requiring accurate measurement.

Equitable principles

While the economic, social, environmental and technical information is being gathered, preliminary formulations should also be made, jointly, of the equitable and legal principles involved. Here international lawyers will be

---

1) C.B. Bourne, "Mediation, Conciliation and Adjudication in the Settlement of International Drainage Basin Disputes", The Canadian Yearbook of International Law, 1971, page 121.

2) Bourne also points out that a single, foreign, technical board may not be trusted by the two countries. Ibid., page 122.

involved. No doubt each country will wish to prepare its case, based on precedents, international declarations, decisions, and treaties. In addition, a joint group should prepare a brief survey of the choices that are open, from the point of view of "rights" to gains and the duty to incur losses.

The main question for exploration here is the "right" of the upstream country. Traditional international law has suggested that when all of a river or lake is within the territory of one country, A, that country has control as complete as over any other part of its territory. Such control would preclude other states, or the community of states, from interfering with A's pollution of the stream.

On the other hand, it may be argued that the full control of each state over its own stretch of the river is limited by a duty to the downstream state. Such a duty might indeed be as complete as an easement (or servitude), independent of the sovereignty over the upstream and downstream regions or industries connected by the river and its flow of pollution. Or it might be a less formidable and enduring duty.

It is significant that while a body of international law has long been accumulating to the effect that a nation contemplating an action on an international river that might harm its neighbours has a duty to confer and negotiate with them with a view to reconciling their conflicts of interest, the approach to transfrontier pollution has been somewhat more demanding. Thus the Helsinki Rules (Article X, para 1(b), 1966) do not merely call for the upstream state to negotiate when its actions may lead to TFP harmful to a downstream state. Instead, they apparently require it to abate existing pollution "... to such an extent that no substantial damage is caused in the territory..." of the downstream state. This positive duty has been stated more officially in the resolutions of the Stockholm Conference (1972). There it is asserted that each nation has a duty to ensure that actions under its control "... do not cause damage to the environment of other states or of areas beyond the limits of national jurisdictions (e. g. the high seas)". (1)

Thus while the general international law of rivers since World War II has been framed in such a way as to impose on a state the duty to negotiate concerning actions deleterious to those downstream; the specific duty concerning pollution has progressed much more vigorously. It might be claimed that such states must now stop any pollution that harms persons or the environment downstream. But enthusiasm for such a doctrine must be tempered by considerations of both fairness and practicability. Does priority matter? For example, should an upstream country that has for years enjoyed an undisputed right now be asked to change its economic structure for the benefit of a downstream environment or industries? Or, does the conduct of the downstream nation matter? For example, should it be allowed to use the same stream to carry its own wastes off to the sea; or should it be compelled to adopt in all its own streams the amount of abatement that its demands are imposing on its upstream neighbour?

Consideration of these questions suggests, correctly, that the causal connection between "duties" and "rights" is very attenuated. Much of the thrust of recent proposals about rights is based upon a Golden Rule maxim: sic utere tuo. Implementation of this maxim seems to call for both states to behave similarly and harmoniously; such an obligation is greatly different from a simple legal finding about "rights".

Thus too much should not be expected from non-political discussions of these unofficial expressions of rights and duties. However, much useful preliminary clearing away of misunderstandings and irrelevancies can be accomplished. (2)

---

1) See C. B. Bramsen, "Transnational Pollution and International Law" in Problems in Transfrontier Pollution, OECD, 1974.
2) The text greatly compresses the subject of case-law on TFP. A fuller discussion would probably begin with the famous Trail Smelter case, stretching between 1928 and 1941. Dealing with the effect and extent in the neighbouring region of the United States of fumes emanating from a smelter in Canada, an International Joint Commission finding recommended that damages be paid and measures taken to abate emissions. This recommendation was eventually confirmed by an arbitral tribunal, which stressed the duty of an upstream state to protect other states.
The text also makes no reference to international law on shared bodies of water such as lakes. The law and proposed law on international rivers also applies to boundary lakes, but most has been said about disputes about the actual position of the boundary. As might be expected, there is more treaty-law about lakes and boundary rivers than about international rivers.

Both principles and treaty-law are changing frequently, and the interests which must be reconciled, along a water-course, are broadening rapidly. As the model in Part II of this paper has made clear, to the upstream state, "rights" to use a river for waste disposal are immensely important, because in subsequent bargaining they can free that state from any obligation to abate, to pay the costs of abating or recompense those downstream who are harmed by TFP. Similarly, to the downstream nation "rights" to receive the waters in their natural state (uncontaminated by TFP) are valuable because in subsequent bargaining it may be able to require abatement, demand payment for TFP, or both. When so much hinges on "rights", no country can be expected to acquiesce in an adverse opinion from a committee of experts. Eventual agreement which compromises conflicting claims is to be expected, supported by other concessions or payments. Nevertheless, the formation of a group of legal experts should assist in posing the leading issues, with authoritative opinions also about the relevance of recent decisions, international conventions, and treaties elsewhere. (1)

The National Impact of Sharing Principles

Beyond these questions of rights, the nations contemplating negotiations will require guidance on the application of such "sharing" principles as Polluter Pays (PPP) and Victim Pays (VPP).

i) Under PPP, the upstream country, A, pays the AC and the downstream country, B, pays the DC. As pointed out above, under this sharing principle there is no compulsion on the part of A to cut back pollution or compensate B for residual damages. Figure 6, panel 2, shows the loss in A's welfare if A is forced to abate. Panel 1 is simply Figure 4 again, and panel 3 shows the impact on B's welfare. (Welfare can be thought of as GNP after all negative and positive differences are taken into account). Panels 2 and 3 are drawn on the assumption that q_A^i is the initial state (welfare positions would be quite different with zero TFP initially). For B, as A is forced to abate under PPP, there is an increase (at a diminishing rate) in welfare until the zero level of pollution is reached. (A's welfare loss is equal to OY in panel 1).

ii) Under VPP, B is responsible for both AC and DC. Through this internalization, B will minimise these total costs and so maximise welfare at q_A^g. Referring back to panel 1, it can be seen that total cost is at a minimum between Q and R at the q_A^g level of pollution. A's welfare stays the same as at q_A^i as AC is paid by B (actually his welfare improves as his internal damages are reduced).

iii) Under CLP (Civil Liability Principle), country A pays both AC and DC and would also minimise cost (and his welfare loss), at q_A^g. B's welfare would immediately jump to the dashed CLP line in panel 3 and stay along it. (This vertical jump is equal to OX in panel 1). The reason is that A compensates for total residual damages.

iv) Under TPLP (Third Party Liability Principle), A is responsible for DC and B is responsible for AC. This is the reverse of PPP. Now, paradoxically, A is motivated to move to zero TFP (loss of welfare at q_A^i is equal to $q_A^i X$ in panel 2 or OX in panel 1) while B is motivated to go to q_A^i (loss of welfare at zero TFP is XZ in panel 3 or OY in panel 1).

v) With the global-rights principle (D'Arge's common property institute principle), the level of pollution would be set by a global institution at q_A^o. B's gain at that point (as if under VPP in panel 3) would be shared with A to offset A's loss (as if under CLP in panel 2) at that point.

vi) Under the status-quo principle, the level of pollution would be set at q_A^i and VPP would be in effect up to that level while CLP would be imposed beyond that level. A's welfare would remain the same as at q_A^i, but B would take the VPP solution and both parties would move back to q_A^o where B would maximise welfare as above.

---

1) The expert preparatory legal group may also be required to survey the rights and duties of the states with respect to other uses of the international river: navigation, consumption, diversion, power generation, storage, drainage, flood prevention, etc. Eventual agreement on TFP may be easier if the parties exchange undertakings to each other, concerning some or all aspects of the basin's development rather than negotiate on TFP alone.

vii) In the dynamic context, A's AC curve rises with economic growth. Despite being forced to compensate B due to the CLP rule, there would be net gains to A's welfare. Figure 7 illustrates the case. In panel 1, after the status-quo principle has been imposed, A's axis origin would shift back by the amount "w" and the old q_A^R would, in essence, become the new q_A^1 (new minimum total cost point for A). In panel 2, B's welfare would rise by OV (due to the gains under VPP) and the same origin shift would occur as for A. Over time, A would experience net gains as economic growth forced the AC curve higher. The locus of the maximum points would describe a net-growth-gain curve, labelled NGG in panel 1. The point q_A^m would describe the maximum amount of pollution A could ever discharge. It would represent the position where the marginal gains from increasing pollution just equalled the marginal damage payments to B. During these growth periods, B's welfare would remain constant along VVq_A^t (the "t" is to remind that time has passed).

If, however, B were the one to experience growth in its pollution sensitive industry or in its polluting industry (i.e. increasing DC curves), there could be net gains in B's welfare. These gains from growth would be indicated by higher VPP curves. The locus of their maximum points describes B's net-growth-gain curve (NGG'). This could lead B to compensate A (for AC) possibly as far back as to the zero TFP level. B's welfare would rise along the NGG' curve up to its maximum, while A's welfare would remain constant along Oq_A^t (leaving aside consideration of less internal damage for A). Both these dynamic cases represent the "he who charges, pays" principle which is perhaps a more accurate designation than the status-quo principle.

viii) Under the agreed level principle, the situation is very similar to the above, status-quo, case. If the level is set at less than q_A^1 the origins will be shifted to show B with more gain than OV and A with less welfare than at q_A^1. If the level is set higher than q_A^1, the origins will be shifted to show B with less gain than OV and A with the same gain as at q_A^1 (until A grows into the higher level, at which time it would show a welfare gain).

CSA: The Incidence of Agreed Shares

We have just seen that the preparatory experts must not only study the decision about the distribution of TFP "rights" and "duties" among the countries, but also the closely related choice among the various financial principles: polluter-pays, victim-pays, third-party-liability, etc. There is still another way to come at this basic decision: the Cost Sharing Approach. Before negotiations can get under way, the negotiators should have fully investigated this approach and the appropriate percentage sharing to be assigned to each country.

In principle, there are two questions to be decided here. The first is the definition, and the determination, of the total amount that is to be shared, and the second is the percentage share to be contributed by each country. There is not much that can be said about either question separately, since countries will agree to one only when they have some idea about the size of the other.

As to the amount to be shared, we may refer to the models in Part II and the remarks on obtaining data earlier in this part. Most writers, speaking of the CSA, have in mind the sum of damage costs and abatement costs in the two countries together. Let us assume that these numbers can be obtained, so that the amount of cost to be shared, as a function of TFP, is known.

What advice can the preparatory group of experts give about the appropriate percentage shares? For example, should country A pay 75% or 15% of the total amount? There are, of course, plenty of precedents in shares of the costs of OECD, the UN, and NATO, in aid to developing countries, and in joint public works and commercial projects. In all of these instances, ad hoc principles have been worked out for dividing the cost and dividing the expected benefits. What has been done before can obviously be done again.

But it is clear that the experts will be unable to recommend a "magic number" to their client nations. Governments are not interested in percentages, but in amounts: each country wants to know what dealing with TFP will cost, compared to whatever domestic alternatives they may have for using the resources or relieving the damages. Hence the experts will be chiefly engaged in compiling a guide to the incidence of various formulae, somewhat as follows.

Each country's share can be thought of as consisting of an initial constant amount, a percentage share of the remaining costs, and a formula by which this share can be itself changed as total costs mount. It will increase the political acceptability if the second component, the percentage share of the remainder, either divides the remainder equally among the countries involved, or divides it in the same proportions as some easily ascertained indicator (such as population, GNP, shoreline, or water-volume (receptive capacity) of the total basin occupied by the countries). (1)

Of course, the choice of the indicator tells the whole story. Some will push most of the incidence on one country, some on the other. If, for example, in the final position (q_A^o) most of the costs are abatement costs, and sharing is according to industrial production, the CSA will in fact approximate to a PPP or civil liability principle: the polluting country will simply be agreeing to cover its own abatement outlays. At the other extreme, if most costs are damage costs, and sharing is according to population, it is quite likely that a densely populated downstream nation will in fact be covering compensation for its own residual damages. Compromise formulae will have intermediate results.

Are such statements all that can be expected from the experts on the incidence of share formulae? There are three other matters that should be considered:

i) Unless the countries are being forced into the agreement by international pressure from outside (or possibly even by threat of armed action) it is obvious that the shares (or the amount) must be set so that every participant is better off under the agreement than it would be without an agreement. This constraint may, in fact, almost completely determine the respective shares. (2)

ii) A country which has already made a substantial effort, especially in abatement might get some initial (or lump sum) credit for this, especially if it can be shown that the abatement effort was for the protection of the other countries, or in anticipation of the agreement. Symmetrically, a country which has suffered unusual amounts, or rapidly-growing TFP, especially during recent periods of protest and negotiation, should get some credit for past damages. While the economist would be inclined to urge that bygones should be bygones, this rule is rather an allocational than a distributional one.

iii) Considerations of social justice may suggest that, ceteris paribus, low shares ought to be awarded to poor and developing nations, and large shares to rich and industrialised nations. Such frequently-cited criteria as equality of sacrifice, for example, point to the acceptability of formulae that pile the burden on those who have a high "ability to pay". Discussion of environmental problems and international relations at the Stockholm Conference confirm a similar doctrine, that environmental efforts should not be undertaken at the expense of, or to slow the growth of, the developing nations.

But such considerations are often simply ill-thought-out slogans. To the extent that they lead to rich countries paying for benefits received by poor countries, they amount to aid or philanthropy in kind; disguised paternalism in fact. Those who urge this kind of sharing should ask themselves whether the beneficiaries would not rather have the help in the form of cash, or industrialization.

This discussion of agreed shares should suggest an additional reason why the CSA approach is unlikely to provide a helpful formula for agreement on TFP. We have seen that in deciding on percentage shares, each country will be forced

---

1) It would be worth-while to study the effects and acceptability of various distributional indicators. For example, see D. W. Pearce, Cost-Benefit Analysis, London, the Macmillan Press Ltd., 1971, pages 26-31.

2) For a discussion of the calculations of countries faced with given shares, see R. Jones, P. H. Pearse, and A. D. Scott, "Joint Selection and Optimization of Projects by Two Jurisdictions", Vancouver, University of British Columbia, (unpublished paper), 1972. Similar points are made in the appendix to the report given pp. 115-175.

to make its own prediction of what the level of pollution, damage and abatement costs are likely to be. Then it must decide the maximum share of total costs it is willing to pay, given the abatement benefit. While one can imagine diplomatic circumstances where making such an involved set of calculations would assist the reaching of agreement, the route seems needlessly circuitous. Furthermore, because they place heavy reliance on early guesses, a safety first psychology may well lead to reluctance to enter into certain types of commitments because of the danger that, when the shared cost aspects are eventually worked out, a country will find itself liable for expenses made by its partner over which it has little or no control. (1)

Strategy

During the preparatory period, as the joint expert groups are preparing agreed background material on the various aspects of possible agreement and joint action, the participating countries will be preparing their own bargaining strategies and positions. They should, of course, be encouraged to do this, and to obtain information as it becomes available, from the joint expert groups. The items which they must be prepared to negotiate will be discussed immediately below, under the heading of "Scope".

It is important, however, that each nation understands and is helped to understand, what it can gain and what it can lose by agreement. Thus, as already discussed, it must know what total costs and its share of these costs are likely to be. Just as important, it must understand what it has to offer the other countries, and what its net benefits will be from going it alone.

i) Bargaining strength - What has a country to offer? The upstream country, of course, has abatement to offer. Our models have mostly implied that the downstream country must offer money payments. If the two countries share a lake or river, they can offer concerted action to each other.

The downstream country can also offer non-cash inducements. Some of them will also be connected with the river or lake: its levels and flows at the border and assistance to international navigation are the most obvious examples. Others may be just as valuable, but will be drawn from the general armoury of international bargaining instruments: trade concessions (suggested by William Baumol); highway, bridge, airport, customs, migration facilities; or reciprocal pollution action in the atmosphere, on rivers that flow the other way, or in a three-way exchange of abatement in a consortium that brings in another country.

ii) Domestic alternatives - What has a country to gain? It cannot know this until it knows what its domestic alternatives are. Some of these are direct alternatives, such as abating pollution that is already TFP, or diluting the flow of TFP by diverting stored water into the river or lake. Others are more indirect, such as compensating the interests that are suffering, or assisting them to adjust by moving to less affected parts of the river. A third group is even more devious, involving direct private payments or bribes to the upstream municipalities and industries that are discharging the pollution; or buying out the offending upstream firms. Finally it is always possible to decide that the return on resources and effort is unlikely to match opportunity costs in quite different domestic projects; this finding might lead to some abatement and compensation coupled with mass migration of industry, recreation and population facilities out of that river basin. (2)

---

1) Furthermore, as Krutilla has shown, changing circumstances may induce a country to reject a particular project which would otherwise be profitable for it. Imagine a joint project which can produce a target X percent return on cost for one country, while still paying the second country its own required opportunity cost of capital. If however, the agreed shares work out so as to return less than X percent to the first country, it will turn to another, inferior, way of obtaining the same benefits by going it alone. In Krutilla's terminology, the country may reject a joint shared cost project if its fixed shares give it a lower benefit-cost ratio than a domestic alternative which it would "pre-empt". John Krutilla, Columbia River Treaty, Johns Hopkins Press, 1967. This defect reflects a well known characteristic of fixed-shares, fixed royalty, or other co-operative arrangements wherein the fixing prevents the owner of an input from obtaining its opportunity cost.

2) We should recall that a country has other foreign alternatives as well. For example, it may induce the UN, NATO, the World Bank or GATT to make international affairs difficult for a recalcitrant neighbour.

Because international-studies textbooks are full of case studies of international bargaining techniques we need not go into the matter further here. Our point is simply that bargaining is both simpler and more likely to produce a high pay-off for the parties if they have done their homework in ascertaining what they have to offer, and what they have to gain.

Scope of Bargaining

Our recurring discussion of CSA shares, and their incidence, introduces the final topic of this section. The preparatory group will assist ultimate bargaining greatly by narrowing down the scope on which agreement must be reached. This task includes the following:

On the one hand it can disabuse the negotiators about the usefulness of certain approaches, provisions, or terms. For example, it can point out that the CSA, per se, will not enable the negotiators to by-pass difficult decisions or precedents about rights and duties (and may actually prevent agreement on certain joint projects). As another example, it can take a view on whether the measurement of damage costs is worth its time and trouble; the group may be able to inform the negotiators that any approach or formula that requires agreement on this statistic is bound to lead to political and nationalistic differences on subjective matters that are insoluble.

On the other hand, it can make sure that all the right subjects are touched on. These include:

1. Agreement on the acceptable amount of TFP by each contaminant, at each season.

2. Procedure in the event of accidental excess of these levels.

3. International payments, transfers, compensation, insurance.

4. Abatement costs, or relevant data.

5. Damage costs, or relevant data.

6. Share of costs, if CSA is to be discussed.

7. Extent of the geographical area considered to contain the abatement and damage costs.

8. Whether other media (other rivers, the atmosphere) are also to be protected.

9. Provision for monitoring and verification.

10. Organisation and finance of an international agency.

11. Through-the-border jurisdiction, or instruments to be used by the other party, or by an agency.

12. Life of the agreement and renewal.

13. Tribunals, appeals, etc.

Bargaining Alternatives

The attached diagram (Figure 9) helps to make clear the alternative outcomes of international bargaining on a TFP situation. After the preliminary expert fact-finding groups have done their job, the nations face a number of choices.

No Agreement

If they do not make an agreement, at least three kinds of outcome may follow. First, both parties can continue to act independently. Each will react to the other's independent TFP behaviour, as suggested earlier in our model of reaction curves. No institutions are needed.

Second, particular TFP situations may be dealt with by suit procedures under public (or private) international law. Even this cannot happen unless there is at least implicit agreement among the countries that certain kinds of suits should be referred to the International Court, or to an arbitrator, mediator or concilator (1). A very limited agreement might also lead to citizens of each using the courts of the other country for TFP suits. The difficulty in establishing individual responsibility in private suits may be reduced if the countries allow themselves to be sued (as Canada did for the Trail Smelter case); but proof of damage must still be offered.

Finally, particular spills, episodes or contingencies can be dealt with as they arise by ad hoc diplomatic procedures. (2) Two countries may agree to join in battling a new pollution problem that affects them both. Or, a downstream country, newly and seriously threatened by TFP, might win immediate agreement by its neighbour to stop this new source pending long term agreement.

It can be seen that each of these require some sort of institutional machinery, perhaps only temporary. It will be obvious that the whole approach of the present paper assumes that such independent action is likely to be inadequate and unsatisfying. However, in addition to the Trail Smelter case and the litigation and arbitration approach in general, it should be pointed out that in marine pollution by oil and by dumping of other wastes, the nations are making progress by a combination of agreement on principles and standards and the empowering of states to detain or arrest ships that are believed to be violating the principles or standards. Compulsory insurance also plays a role. This whole type of approach is (necessarily) different from that most appropriate for TFP of rivers, but may have some lessons for dealing with contingencies.

The Swiss Corporation

If the parties do proceed to a substantial and complex agreement on TFP, a number of possible institutional forms may emerge (see Figure 9). One of these is a specialized international agency of the Swiss Corporation type. This would be an operating agency that in principle would "sell" abating services and "compensate" for damages, attempting as far as possible to maximize its costs. (3) The owners of its equity would be the two or more countries concerned with TFP.

The advantage of such an institution (which would appear to the people living along the river to be a public utility) would be that it could ignore the border, internalizing in a hard-headed manner the choice between abatement and damages.

Its singleness of purpose, however, would be weakened by two problems.

In the first place, it would be little better able to cope with the measurement of damages than any alternative institution. Thus on the one hand it would, as a non-political body, have great difficulty in deciding how much damage to prevent. Indeed it is frequently argued by public finance theorists that a sufficient reason for the existence of governments is precisely the political determination of the amount of provision of such public goods as damage reduction.

Three interesting counter arguments to this pessimism have been offered. All suggest that the Corporation might be even better situated than national governments to obtain measurements of damage. The first is that the Corporation should use the courts of the downstream country to set the amount of damage or expropriation cases. It would then be up to the Corporation to decide whether to pay this amount as compensation for residual damages, or whether to reduce such expenses by undertaking upstream pollution abatement instead. The requirements of such a judicial role would clearly be somewhat different from those of today's municipal courts. Further, to discover the "public goods" values of certain river

1) Cf. C. B. Bourne, op. cit. pages 138-151, for a full discussion.

2) Diplomatic procedure might also be adequate for eventually obtaining simple, parallel, harmonious, national pollution legislation, independent of TFP. This appears to be the aim of the Council of Europe's draft legislation on air pollution. See Klaus Boisseree, "Chances and Problems of International Agreements on Environmental Pollution", Natural Resources Journal, 12, 2, April 1972, pages 218-226.

3) One version of this institution's mode of action has been extensively worked out by Smets, in the context of a discussion on the principle of "mutual compensation". (See report given pp. 115-175)

uses threatened by TFP, or to learn the private values of certain non-marketed recreation or aesthetic pleasures, where property institutions are conspicuous by their absence, would produce more difficulties. And the Corporation would get little help from the courts in learning the long term values of damage functions. (e. g., the smaller damage that would remain when polluters and victims had migrated from their present locations to more protected areas or areas with higher assimilative capacity). Nevertheless, it can be claimed that however inept the courts alone might be in thus determining damage functions adequate for both efficient allocation and equitable sharing, the Corporation would be a more receptive and acquiescent "defendant" than a government department. Indeed the Corporation and the courts, between them, might be able to accomplish damage measurement where a government would fail.

It can also be argued that the Corporation could use local (and provincial) governments to ascertain damage costs. For example, the Corporation could announce that it would accept offers from localities to abate the pollution from which they suffered. Analytically, such offers would be roughly similar to those local governments make in buying supplies and labour services for the provision of local public goods and indentical to offers (or grants or subsidies) to induce firms or neighbouring jurisdictions to take actions that provide desired spillovers or external benefits. Indeed, compared with national governments, the Swiss Corporation would have the advantage of having narrow powers so that localities would have little incentive to make other than straightforward offers for damage abatement. (Similar and symmetrical remarks would apply to the offer to sell abatement services by upstream localities and firms).

However, if more than one downstream locality would enjoy abatement simultaneously, disclosure would become a difficulty. All would perceive the opportunity to be "free riders" on the coattails of other jurisdictions seeking abatement. Even here a Corporation might do better than a senior government, however, in arriving at some modus vivendi by which it could ultimately discover the whole amount of downstream damage.

Whether a locality attempts to be a "free rider" depends on whether there are many or a few such downstream jurisdictions. This is touched on in the Smets outline of one version of the Swiss Corporation, and also in the D'Arge account of a common property resource institution. The downstream localities, if they were few enough, would not attempt to be free riders; they would be more like club members. They would perceive that if they overstated the amount of damage, they would be provided with, and have to pay for, too much abatement. Thus a Swiss Corporation, dealing with only a few downstream clients, might be able to overcome some of the difficulties of ignorance about damages, by taking account of the capacity of local jurisdictions to gauge the communal willingness-to-pay of their residents. In this they would be better situated than national governments, who would find it impossible to deal dispassionately, impersonally or non-politically with local authorities.

A second difficulty about the Swiss Corporation would be its propensity to take advantage of its monopoly position (perhaps unconsciously). Ideally, if the frontier is to be ignored, the Corporation's terms of reference should instruct it to equate marginal damage costs with marginal abatement costs, which is the same as minimising the sum of upstream and downstream AC and DC. It would be much more likely, however, to take account of its own effect on the "price" it had to pay for abatement, and the "price" it was offered for damage reduction. If abatement costs were constant and the damage curve were discontinuous at some "standard" level of pollution, this monopolistic propensity would make no difference to the amount of abatement the Corporation decided to carry out. But, as is more probable, if the marginal damage and abatement cost curves were increasing, then the Corporation's abating activities would become distorted, and confused with the ability to make monopoly profits. If the Corporation were selling waste disposal services to firms, it would tend to sell too little pollution, with damages less than marginal abatement costs. Just as serious, if the Corporation were selling abatement services, it would tend to offer too little abatement, with marginal damage well above the marginal abatement costs. Whether this problem would be serious in practice is a matter of fact, not necessarily a significant disadvantage of the Swiss Corporation scheme. (1)

1) This problem seems to be similar to various problems described by R. C. D'Arge and A. V. Kneese, "The Economics of State Responsibility and

Cost sharing agreements

The second form of agreement would involve trying to implement the cost sharing approach (CSA). Here the countries would make prior agreement on the costs to be included in the total, and the percentage shares of this amount to be born by each country. They would then experiment with various amounts of abatement, until the total share borne by each was minimised.

We have already come across various objections to this approach. From a bargaining point of view, the scheme would also seem to be very impractical. Even after prior agreement on shares and definitions had been reached, the countries would somehow have to set up an institution or agency within which the results of the experiments with different levels of costs could be transformed into decisions about the total amount of abatement (and damage). The two nations would entrust it not only with verification of costs and monitoring of TFP, but also the decision about which level of TFP carried the minimum total cost. (If they did not do so, but examined each level of cost themselves, then re-negotiated the fixed cost shares, the approach would no longer be faithful to the CSA idea). The countries cannot act without an "agency" because the downstream country cannot reach into the other country to control the amount of abatement, nor will it simply hand it a sum of money, saying "spend that". Reflection on this role suggests that the "agency", or political council, that would thus be responsible for monitoring and decision making would in fact have to be unusually autonomous.

Such a task would not be unusual for an inter-governmental agency; it would be less independent than a "Swiss Corporation". But it seems quite unlikely that two nations that were unable to agree on a level of TFP, plus transfer payments (and so had recourse to a CSA), would turn out to be more willing to grant so much autonomy to a TFP agency.

Negotiated TFP level

The third, and most promising, form of agreement would result from a triple bargain: the level of TFP at the border (also called a quota); the payments in cash or kind to be made among the countries; and the institutions to accompany or administer this decision (see Figure 9).

The table makes clear that this third approach is most promising partly because it is least specific. Along the left hand edge are components of any negotiated treaty on TFP, and the columns in the table show four sets or groupings on entries of A, X or O, that together indicate a particular style or type of agreement. Going from left to right, in the table, indicates an increasing degree of delegated authority to a treaty organisation, or agency. If the agreement makes provision for a certain role to be undertaken at all - either by the agency of by the governments working together in some other way - that provision is indicated by entering an A (agency) or X (other). A zero, O, is entered if the agreement makes no provision for that role or function. Similar entries are also provided, separately, for the CSA, as already discussed.

There will, of course, be scope in any agreement for more subjects than these. Because it deals with the use and improvement of a river basin or watershed, for example, it might record as well settlement of other transfrontier-watershed matters not directly related to TFP: navigation, power production, levels and flows, flood protection, and so forth. If an agency were to be set up by the agreement, for example, it might be convenient to use its accumulating experience and prestige to tackle similar questions. Or, if the quid pro quo for an upstream-downstream bargain were to consist of undertakings or concessions in the opposite direction, it might be incumbent to provide for these in the same treaty - for visibility, as it were. It is appropriate to warn countries against loading down a single treaty organisation with too many functions; much is to be gained by having several such bodies, if they can keep their functions disentangled.

Note 1) (Cont'd):

Liability for Environment Degradation", Riverside, University of California (unpublished draft prepared for the American Society of International Law), 1973; and Henri Smets in various papers, where systems of payments would not balance, where profits or losses would slowly accumulate, or where "efficiency" would not be achieved.

These considerations lead us to ask, what criteria should we adopt in choosing a TFP solution or institution? We have so far stressed real costs (AC and DC) and their separate and joint minimisation. In addition to these, we must keep other, more nebulous, cost concepts in mind: the real concepts of transactions between jurisdictions and agencies; the costs of agreement or consent within a jurisdiction or agency; and the costs of compliance imposed on citizens or firms by an agreement or its administration.

i) As a base case, for comparison, we could consider a TFP agreement in which no agency is set up. Previous sections have discussed the countries' progress toward agreement on a norm or level of TFP at the boundary, carefully defined as to mode of measurement, seasonal variation, allowable error, and so forth, for each pollutant. This agreed level must also have a time dimension: is it to be the same every year, or is there a schedule or formula showing a program of annual reductions (or increases, if economic development points that way)? Finally, they must decide how to handle pollutants not specifically provided for (new chemicals, for example), the duration of the agreement, and the mode of revision, renewal or denunciation.

The countries should then, at a minimum, have a meeting of minds on the items listed along the left hand edge of the table. Monitoring of water flows and water quality will be necessary, as will verification (or auditing) of physical economic and social data supplied by the countries to each other. If payments or some quid pro quo in kind are called for, the mode and timing of payment should be specified.

In this simple base case, the countries will undoubtedly leave to each other's independent jurisdiction the tasks of constructing and financing abatement works, operating them, and relations with private origins and victims of pollution (using regulation, subsidy or charge).

They will also have to make decisions about "contingencies" and appeals. Contingencies differ from the sort of TFP we have discussed so far in that they are unexpected, arising from natural causes or human error in handling materials and wastes. Spills from refineries, chemical tanks, ships and waste-abatement plants are good (or bad) examples. When such spills take place, two series of events must proceed rapidly and smoothly, both calling for preparation and training.

On the one hand, because there will eventually be an investigation possibly leading to compensation of or insurance awards to the victims, penalties on the sources, or revision of the agreement, (or all of these), it will be necessary to fully record everything that happens. Monitoring should become more frequent and accurate, and statement of witnesses noted. Suspected persons or ships who might be tempted to leave the region should be prevented from doing so. An accounting should be made of amounts paid for emergency relief of those whose homes or plants are damaged.

On the other hand, the countries should have a rehearsed emergency procedure for dealing with the spill itself. Very often this will require equipment, experts and manpower from both countries. The amounts of such contributions, the leadership or coordination of the contingency forces, and their powers to order, and to conscript fully investigated and understood. The role of insurers must be decided.

As part of the general provisions of the base case, the countries must have also an understanding about appeals. These will be both private and public. Are private suits or appeals to be permitted or encouraged, or should all TFP items be handled by the national governments? In addition to private complaints, of course, the governments themselves will not always be satisfied with the performance of each other. They may have technical differences over the interpretation of the agreement with respect to levels and changes in TFP, or they may distrust current reports about abatement or damages. And they may differ about the amounts, timing or payment of financial and in-kind transfers. How should such differences be resolved? At one extreme, diplomatic channels may be used. But it is more likely that provision will be made for conciliation, mediation and arbitration, including some independent investigation. (1) Even if no other agency

---

1) This is the subject of C.B. Bourne "Mediation, Conciliation and Adjudication in the Settlement of International Drainage Basin Disputes", Canadian Yearbook of International Law, 1971, pages 114-158, and there is substantial legal literature on dispute settlement, both within and outside existing treaty arrangements.

is created by the agreement, the countries should consider creating a standing commission or tribunal for hearing appeals on technical matters, for nothing is more important to action on TFP than **joint** acceptance of otherwise disputed "facts".

Finally, the nations should make provision for changing circumstances. It is to be presumed that technical and taste changes will lead to unforeseen demands and opportunities for improving the agreement. Should these be ignored during the life of the treaty; and should that life be made correspondingly short so that revision may be frequent? Or should monitoring and verification be supplemented by "research" leading to revision during the life of the treaty (for changed agreed levels, for example)? What arguments should be made to assure adequate, acceptable research findings, and for revision on the basis of these findings?

The preceding paragraphs have given some idea of the meaning of the required components of any agreement. The base case has been one in which the agreement is assumed not to have involved much, or any delegated autonomy. No special TFP agency has been set up. Consequently after the costs of initial international information and agreement have been incurred, the nations will have two other kinds of costs. First, their governments must impose costs of transactions or consent on their own persons or firms, as they intermediate between them and the other nation. To the people, such costs will not be associated with TFP, but will simply be the costs of compliance with domestic laws and regulations. Second, the governments will have continuing costs of arms-length negotiations on technical TFP decisions with the other government. The lack of a powerful international agency may help to preserve autonomy, but only at the price of very expensive, perpetual negotiation.

Such a base case would improve only very desperate situations. Its lack of organisation or coherence is particularly deficient in that it does not create new means by which accepted information could be generated, transmitted or considered. Costs are kept low because some functions are not performed, and others are not perceived.

It cannot be too often urged that what is required is an agency or commission. Its first and principle function would be the obtaining and interpretation of data on TFP. It would be manned by competent personnel from both countries, and should be adequate to assure that differences of opinion about the trends, levels and so on could not exist. This role is usually referred to as "monitoring", but that word may not create the right impression. It is not necessary that the agency actually measure TFP with its own staff and equipment. Economy may dictate otherwise. The agency must simply be able to assure both countries that it has received information from monitoring, irregular sampling, visual inspection, and so forth; has verified any part of the information that might be unrepresentative or subject to bias on the part of those collecting it; and will full consent of recognised, trained, objective personnel from both countries, come to certain conclusions, preferably states numerically.

The non-technical, lay or political commissioners or members of the agency will then be placed in the position where they can (in interpreting problems and performance in the light of treaty obligations and schedules) come to unanimous conclusions and recommendations. This in turn will prevent the agency from degenerating into a distrusted forum for partisan debate, but keep it as a respected, international (non-national) body to which the nations are increasingly willing to entrust minor decisions. (1)

In addition to this essential role, the agency might be entrusted with other tasks. In the table, it has been suggested that the coordinating of contingency

---

1) See A. D. P. Heeney, Along the Common Frontier - The International Joint Commission, (Canadian Institute of International Affairs Behind the Headlines Series), Toronto Vol. XXVI, 5 July 1967, for a discussion of the IJC's expert boards. Many other papers and speeches about the IJC make similar points. The bi-national, official, composition of IJC boards of course has advantages in economy and flexibility, and also disadvantages in the possibility of some loss of control. But its great merit is that the two national sections of an expert board "always come together to report as one to the Commission and any separate investigations they undertake are planned so as to complement each other". Heeney, page 9.

preparations might be particularly suitable. The composition of the agency might also indicate its availability to undertake research leading to revision, or, quite different, to hear appeals and act as a private case tribunal.

In essence, however, this would be a mini-agency. It would not need powers over enforcement or payment to do its own job well, and it could easily co-exist with a system of separate but coordinated performance by national agencies of the other roles in the table, or indeed with other international agencies in the river basin.

ii) Before turning to more elaborate, autonomous agencies, we may examine here the certificates scheme (1). Although this scheme has several distinct advantages, cost saving is not one of them. It does not enable the countries to dispense with the information gathering and negotiation phases of the other types of agreement. In most variants, the need for data collection before negotiation begins, and for informed bargaining (and even strategic behaviour) is still present. For, just as with the "agency" schemes sketched later, there must eventually be consent to a target level of TFP, a timetable for reaching it, and some understanding about the sharing of the costs. Indeed, in the original version of the scheme, it was suggested that a main advantage was that it forced participating nations to think in advance about the list of considerations that might otherwise be neglected in negotiations. Its preliminary costs, therefore, are likely to be at least as high as with rival schemes. Furthermore, its continued efficient operation requires that many persons, firms and governments operate a market, with all the information, decision and transactions costs that that implies.

What then are the more positive advantages of the certificates approach? There are two: efficiency and flexibility. Given the initial decisions about rights and time schedules, the certificates scheme tends automatically to assign permitted pollution (near to q_A^0 in our model) to those polluters whose alternatives - that is abatement costs - are the highest, regardless of country. Thus the cost of abating to any given level of TFP is minimised.

The flexibility flows from a fairly mechanical extention of the scheme to changing circumstances. (It has already been anticipated as the status quo, or, he-who-changes-pays, principle). Even though an initial level of TFP has been agreed upon, it is possible for new polluters to get agreement to increased discharges by buying more certificates from the victims or from other polluters. Similarly, it is possible for the victims to win a lower level of TFP by buying more certificates from the polluters. Thus changes from the set level of TFP would be at the expense of those who desire the change. (2)

---

1) First suggested in C. B. Bramsen and Anthony Scott, "Draft Guiding Principles Concerning Transfrontier Pollution", Problems in Transfrontier Pollution, Paris, OECD, 1974. Comments on this scheme are to be found in the papers by Henri Smets and particularly in the most recent April 1973 paper, where with certain minor alterations it is viewed as a means of embodying the Mutual Compensation principle. See report pp. 115-176.

2) Notes on the certificate scheme: (a) In the original presentation, it was implied that the polluter paid for the initial change from q_A^i to q_A^0. This PPP is not a necessary feature: there can easily be provision for an initial transfer to the upstream country; or the initial change can be a small one, with all further abatement paid for by the downstream buying of rights. Indeed, any distributional feature can be built in. (b) It was originally proposed that the transition from q_A^i to q_A^0 be handled by having the certificates themselves "depreciate" in the amount of pollution right they conveyed. This still seems like a good idea, but it is not essential. One or both countries could buy in certificates every year, or rights could be re-auctioned, in smaller amounts every two or three years. (c) The downstream country, or its provinces or municipalities, would be the ideal downstream holders of the certificates. Since much abatement produces "public good" benefits, it is unlikely that their value would be very high to individuals (or even firms) enjoying abatement. This characteristic of the scheme need not be specially provided for, because there is no chance of downstream individuals wishing to hold certificates. But non-governmental groups (fishing, environmental and recreational) could well be large enough to benefit from purchasing certificates. However, in general, we would expect the downstream holding of rights to become concentrated, and therefore less competitive. (d) It has been objected that the market in certificates is unlikely to be a good allocator of pollution rights, because large firms and governments are

The footnote to the previous paragraph goes into some of the details of a certificate scheme. But the outlines are simple enough: the scheme forces forethought on the bargainers; it leads to efficiency in abatement, and it is highly flexible in the face of changes in taste, industrial structure, location or sheer size. Once the initial decisions have been made, the costs of changes in administration and transactions would devolve chiefly on persons, firms and junior governments. There is any amount of scope for senior governments to act, if they wish; but if they do not, the system can become "cosmopolitan", sensitive to geography but not to boundary lines.

iii) From the certificates scheme we turn to the characteristics of schemes that center on agencies and super-agencies. The only difference between these is the number of functions delegated to them. A super-agency might well become so complex and so autonomous as to become an international decision maker for many aspects (location, employment, taxes, and industrial structure) of the lives of the peoples grouped together in the river basin. But an agency would normally accumulate only a few tasks beyond those sketched out in our base case a few pages earlier, and make few decisions without reference to its parent governments and their departments.

Three points should be made, having to do with through-the-border controls, distributional decisions, and costs. It is often urged that international agencies should have direct powers to impose their controls on the nationals of the participating nations. Three examples come to mind. (a) In some international fishery conventions, "observers" are placed on board each vessel. These observers are therefore a sort of inspector or policeman working for the convention management itself. (b) Within the United States, and in some other federations, there is some willingness to let inter-state agencies (set up by the states) take over the jurisdiction of the members over certain matters. (c) In the European common market, and in the Pan-European movement, there is continuing agitation in favour of governments relinquishing their decision, or administrative powers, to some super-national agency. It is not the economist's task to discuss the pros and cons of this idea as it is applied to TFP. The ramifications of principle, and of convenience, autonomy, sovereignty, consent, and administrative feasibility stretch far beyond the few principles on which this essay has been based. We are here forced simply to lump them together as "transactions" and "agreement" costs.

Note 1) (Cont'd):

likely to dominate and manipulate the "market". Such firms have both a large demand and wealth, so that the scheme will not necessarily allocate the rights to those whose abatement costs are highest. In rebuttal, we should note that there are actually two points here. It is true that if the participants are very few in number, certificate pricing will not be competitive and the resulting allocation far from perfect. But will it be worse than alternative schemes? Will the rights gravitate to wealthy corporations from the poor man? I do not believe they will. The issue already comes up in the holding of quota certificates for raw materials, for tariff schemes, fisheries, etc. While one finds that wealthier participants need and demand more certificates there is no evidence that their sheer wealth or size inclines them to carelessly offer more than small participants. While imperfection in the capital market may impede certificate acquisition by those whose access to funds is most expensive, the persistence of land-holding by small farmers, and of share-holding by small investors, suggests that market imperfections do not necessarily lead to greater concentration than is already to be found in the distribution of productive assets. With respect to non-industrial (domestic) holding of pollution rights, two opposite outcomes can be inspected. At one extreme, the would-be polluters may be farmers, householders and summer people. Along most international rivers, such persons would be already relatively wealthy, and quite able to get access to capital markets to buy certificates (or to abate). At the other extreme, however, the riparians might be low-income people in villages and towns. They might well be unable to compete, even by collective borrowing, for certificates. If so, good social policy would indicate public intervention on their behalf. (e) Certificates are not simple things. There would have to be one issue for each kind of pollutant. Furthermore, if the downstream effect of a pollutant depends upon the location of the discharge along the river, each certificate would have to be location-specific. In both cases the experts might be able to determine the relative pollution value of certificates as they were traded between different places or materials.

All that needs to be said here is that it is possible the countries would agree to imposing regulations, taxes or subsidies from some central authority onto firms, municipalities and persons in both countries. It is usually assumed that these measures (instruments) would be uniform, applying equally to persons in both countries. (1) But even this constraint is not essential. Once the general principles of an agreement have been accepted, the agency may be authorized to proceed to carry it out, including measures that bear unequally on the different locations, persons or firms under its jurisdiction.

Distributional changes: in some super-agencies it may be desired to assign so much authority to the agency that it feels free to change the levels, or the payments, by consent of its members, without more than the power of veto being left to the parent governments. Once again, these may change the balance of benefits between the two countries. It would be hard to build an effective super-agency if it did not have such powers; but it would be an unusual agency if it did. In the only important examples, fishery conventions, the negotiation of catch quotas among the member nations is not carried on within the agency, but separately, among the nations meeting "in another place". In general, therefore, one would expect that both through-the-border powers and distributional changes would rarely be assigned to agencies; they would depend on national governments (and further negotiations) for these decisions. Indeed, the objectivity and credibility of the agency is likely to be damaged to the extent that it is in a position to act "unfairly" to the citizens of some country.

What would be the costs of a complex agency? The time and trouble of persons acting within it can easily be calculated, but what about the member governments and their citizens?

In my opinion, an agency cannot be justified a priori on the basis of administrative economies. Indeed it sets up new channels and new obstacles to common sense cooperation at various levels. Agreement, administration, and communication may actually become more expensive. Not only must the governments bargain among themselves, but they must now bargain with the new super-entity itself. There is no reason to believe that the agency will be a costly device, for it may equally well (as a clearing house, and as a trusted, objective, institution) make the diplomatic relations of the two countries easier. Just stating the sources of new costs and new economies cannot settle the matter. The real gain from an agency is that, by providing channels through which agreed and certain information is presented in accustomed form to experienced decision makers, it may enable different and more efficient decisions to be made. While the calculus of transactions and agreement costs cannot be solved without data applicable to each case, it can be predicted that the presence of some objective, shared, agency will enable the countries to at least consider both joint action and reciprocal action-and-payments. The consequent tremendous reduction in ignorance, uncertainty and partisanship will inevitably make the countries better off, regardless of the degree to which they decide to reduce their day-to-day autonomy.

---

1) For one discussion see J. M. Cumberland, "The Role of Uniform Standards in International Environmental Management", Problems of Environmental Economies, Paris, OECD, 1972, pages 239-253. Papers by D'Arge and others seem to contemplate similar ideas. The political science literature is chiefly about the functional agencies of the UN, and so does not deal with bilateral functionalism.

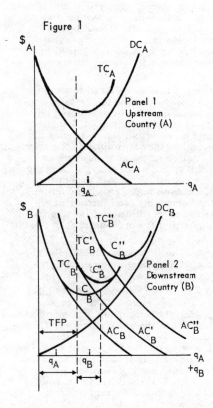

Figure 1

$A

DC_A

TC_A

Panel 1
Upstream
Country (A)

AC_A

$_A$

q_A

$B

TC''_B   DC_B

TC'_B   C''_B

TC_B   C'_B

C_B

Panel 2
Downstream
Country (B)

TFP

AC_B   AC'_B   AC''_B

q_A q_B q_A
+q_B

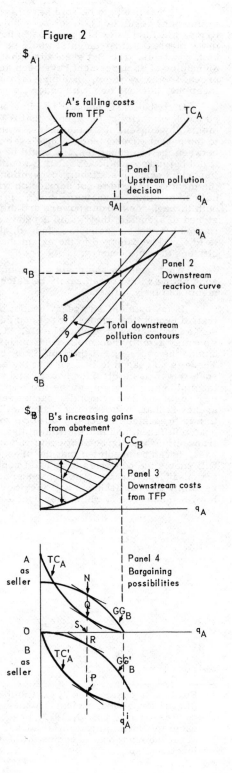

Figure 2

$A

A's falling costs
from TFP

TC_A

Panel 1
Upstream pollution
decision

q_A

q_A

q_B

Panel 2
Downstream
reaction curve

8
9
10

Total downstream
pollution contours

q_B

$_B$ B's increasing gains
from abatement

CC_B

Panel 3
Downstream costs
from TFP

q_A

A
as
seller

TC_A

N

Q

GG_B

Panel 4
Bargaining
possibilities

O

S

R

q_A

B
as
seller

TC'_A

P

GG'_B

q_A

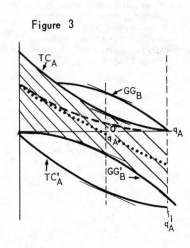

Figure 3

TC_A

GG_B

O

q_A

q_A

TC'_A

GG'_B

q_A

216

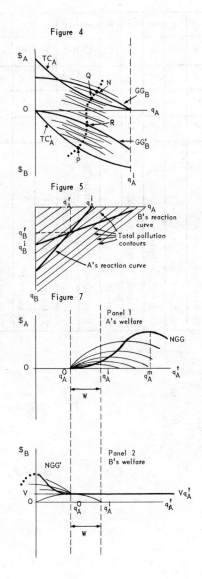

Figure 4

Figure 5

Figure 7

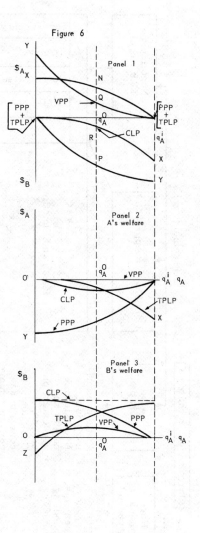

Figure 6

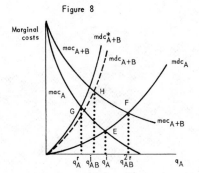

Figure 8

CLP   = Civil liability principle.
VPP   = Victim pays principle.
PPP   = Polluter pays principle.
TPLP = Third party liability principle.

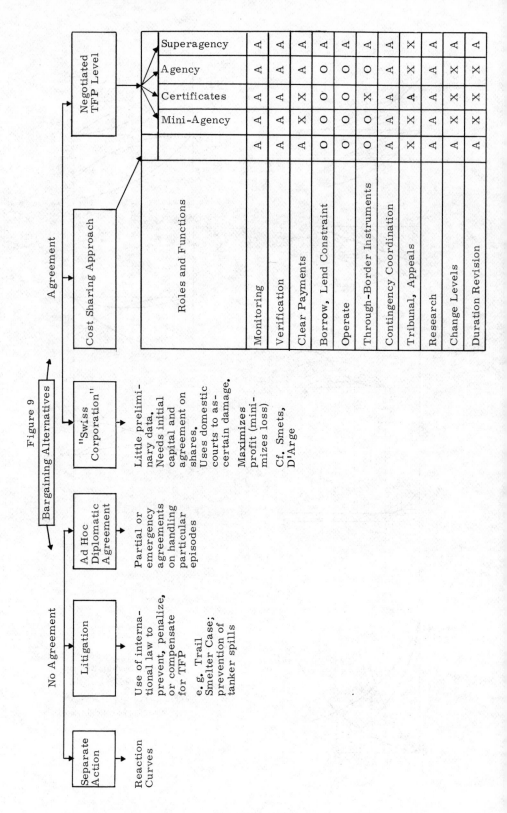

Figure 9